THE PURSUIT
OF VICTORY

THE PURSUIT OF VICTORY

From Napoleon to Saddam Hussein

Brian Bond

OXFORD UNIVERSITY PRESS

1996

Oxford University Press, Walton Street, Oxford OX2 6DP

Oxford New York
Athens Auckland Bangkok Bombay
Calcutta Cape Town Dar es Salaam Delhi
Florence Hong Kong Istanbul Karachi
Kuala Lumpur Madras Madrid Melbourne
Mexico City Nairobi Paris Singapore
Taipei Tokyo Toronto
and associated companies in
Berlin Ibadan

Oxford is a trade mark of Oxford University Press

Published in the United States
by Oxford University Press Inc. New York

British Library Cataloguing in Publication Data
Data available

Library of Congress Cataloging in Publication Data
Bond, Brian.
The pursuit of victory: from Napoleon to Saddam Hussein/Brian
Bond.
p. cm.
Includes bibliographical references and index.
1. Military history, Modern. I. Title.
D214.B66 1996
327.1'17'0903—dc20 95–16013

ISBN 0–19–820497–3
1 3 5 7 9 10 8 6 4 2

Typeset by Best-set Typesetter Ltd., Hong Kong
Printed in Great Britain
on acid-free paper by
Bookcraft Ltd., Midsomer Norton, Bath

ACKNOWLEDGEMENTS

The idea for the theme of this book occurred to me during a telephone conversation with the military historian Richard Holmes, to whom I am most grateful for suggesting a broad and surprisingly neglected subject on which I could work during my sabbatical year. Those discouraging individuals who advised me the subject had 'already been done' by an author whose name they could not recall shall remain nameless, whereas Sir Michael Howard cheeringly reassured me that 'nothing has been done'.

During the academic year 1992/3 I was a Visiting Fellow at Brasenose College Oxford, and I am glad to take this opportunity to thank the Principal and Fellows for many stimulating discussions over lunch, not all of them about the history of warfare and the progress of my research. In the Michaelmas term I helped to organize and chair a research seminar at All Souls College where Robert O'Neill kindly arranged for me to use the Codrington Library. The well-stocked military history library at King's College London was also invaluable.

I particularly wish to thank Christopher Duffy, Jan Willem Honig, and Lawrence Freedman for giving me the benefit of their expertise in commenting on individual chapters, and above all to thank Brian Holden Reid for his detailed constructive comments on the whole draft: I hope he will understand why I have not devoted more space to the American Civil War. I am grateful to all the above, and other colleagues, for lending me books and suggesting others which I should consult. I have accepted and incorporated nearly all their corrections and criticisms, but of course I remain entirely responsible for any surviving imperfections.

Most of the book was written at home during my sabbatical, free from all distraction except for televised sport, fox- and badger-watching, and the lure of the garden. My wife Madeleine not only supervised this comfortable and well-provisioned haven, but also typed and retyped the whole draft from my neat but illegible longhand.

My editor at Oxford University Press, Tony Morris, has supplied just the kind of support, encouragement, and patience which every author needs.

Lastly, I should like to thank Andrew Orgill who has once again contributed an excellent index.

                                                                    B.B.

Medmenham, Bucks.
September 1994

CONTENTS

LIST OF PLATES

(between pp.118 and 119)

LIST OF MAPS

Acknowledgements to the maps

Maps 1, 2, 3, and 5 are based on maps kindly supplied by David G. Chandler from the *Atlas of Military Strategy* (Arms and Armour Press, 1980). Maps 4 and 6 are based on maps by W. Carr from *The German Wars of Unification* (Longman, 1991). Maps 7 to 13 are based on maps by M. Gilbert from the *First World War Atlas* (Routledge, 1970), from *The Atlas of Recent History* (Routledge, 1967) and the *Jewish History Atlas* (Routledge, 1969).

❧ *Introduction* ❧

'And everybody praised the Duke,
Who this great fight did win'.
'But what good came of it at last?'
Quoth little Peterkin:
'Why, that I cannot tell', said he,
'But 'twas a famous victory'.

(Robert Southey, *After Blenheim*)

From the viewpoint of its victims and their descendants, whether at Blenheim, Waterloo, Sedan, or Stalingrad, every victory is likely to be incomprehensible and hollow, if not indeed in the words of little Peterkin's sister 'a very wicked thing'. Their grandparents had been expelled from the battlefield and their home burnt; the country round 'with fire and sword was wasted far and wide'.

But Southey's bitter verses are suffused with irony: he knew (and we know) that terrible and destructive as war is, victory is usually sharply differentiated from defeat and often has profound and long-lasting political effects. Successful wars have played a vital role in establishing states and enhancing their power. Conversely, defeats have frequently led to widespread destruction, harsh occupation, and humiliation of the people—fates which Britain, the former Dominions, and the United States have been spared in modern history.

But what constitutes victory? All students of history must be struck by the ambivalence, irony, or transience of most military victories, however spectacular and 'decisive' they appear at the time. Yet apart from the enduring appeal of volumes devoted to 'decisive battles' and 'decisive wars' (discussed further in Chapter 3), comparatively few historians have focused on the issues of victory and defeat in war. The present study consequently seeks to open up an interesting and significant topic for further research and analysis.

In classical warfare, battles seldom lasted for longer than a single day and were frequently 'decisive' in the sense that one side was defeated on the field, harried and slaughtered while retreating, and deprived of any capacity to offer further resistance. With the development of the modern states system in Europe from the mid-seventeenth century such rapid, decisive victories became harder to achieve. Wars would normally be decided over more than one campaigning season, in which the eventual result would be achieved by cumulative tactical victories accompanied by the attrition of manpower, military stores, food supplies, and financial reserves. In a recent book, Russell Weigley characterizes the period 1632–1815 as *The Age Of Battles*,[1] but deplores the fact that wars were seldom decided in a single day and were therefore excessively costly. He understates the interplay of military and political skills which could guide individual states or alliances to eventual success in war and then reap the benefit (usually measured in territorial gains) in the post-war settlements.

The present study begins with the dynastic or 'Cabinet' wars of the mid-eighteenth century where nationalist or religious fervour was usually absent, and the bulk of the populations of the warring states were not deeply involved. Frederick the Great of Prussia is selected as the outstanding soldier-statesman of the age, who resorted to war with cynical realism—abrogating treaties and invading neighbouring provinces without a declaration of war—in a successful attempt to enhance his scattered and impoverished kingdom's territory and power. The benefits or dividends of victory in land (and perhaps even more sea warfare) in this period were clearly manifest in the peace treaties. There was plenty of 'slack' in the European states system in the form of principalities, smaller territories, and even individual fortresses which frequently changed hands, and in addition vast overseas colonies to be competed for by the maritime powers. The principal European states waxed and waned in size and power within widely understood conventions, but seldom sought the complete destruction of their rivals and fellow actors in the dynastic game. The Prussian state might have been eliminated had Frederick been decisively beaten, but the repeated divisions and eventual extinction of the state of Poland in the 1790s was, nevertheless, exceptional.

French revolutionary warfare, pushed to its extreme form by the military genius of Napoleon, embodied a sharp contrast to the more limited operations of the preceding era. Napoleon, in particular, exploited revolutionary fervour and mass mobilization to the full. He pursued grandiose political goals and adopted a dramatically offensive style of operations, characterized by rapid grand-scale movements and a new stress on decisive battle as the planned culmination of the whole campaign. Combining the roles of emperor and commander-in-chief, Napoleon's most brilliant victories (including Ulm, Austerlitz, and Jena) resulted in the temporary crushing of his opponents' armies and their acceptance of punitive peace terms. At the peak of his success, *c*.1805–7, Napoleon had conquered most of Europe and made his conquests pay in terms of revenues, advantageous trading arrangements, and the provision of military contingents. In doing this, however, he put himself on a treadmill of perpetual wars to keep France and its immense military machine solvent. As a statesman he conspicuously failed to retain allies, conciliate opponents, or secure a general settlement while still winning. He was eventually beaten, first in 1813–14, and finally in 1815, by a coalition which managed to remain united long enough to overwhelm him by weight of numbers. Waterloo was a 'decisive battle' by almost any criterion.

So far from his reputation as a military genius being eclipsed by eventual defeat, the Napoleonic myth of a heroic quest for glory through military victory—obtained by the destruction of the enemy in a decisive battle— dominated European military thought right up to the First World War.[2] The Napoleonic style of war was promulgated as the ideal to strive for by numerous historians and theorists, of whom two were outstanding in terms of the quantity, quality, and influence of their publications; namely the Swiss soldier of fortune Antoine Henri Jomini and the Prussian career officer, Carl von Clausewitz. Jomini's crucial importance for the present study is that he discerned permanently valid strategic principles from his studies of Napoleonic warfare which he inculcated with clarity and immense confidence through his own long lifetime (he died in 1869), and thereafter indirectly for several generations through devoted disciples or followers in both Europe and the United States. If any one writer can be said to have shaped the military mind up to and beyond the First World War, as regards the assimilation of a view of generalship based on the application of objective strategic principles, it was surely Jomini.

By contrast, Clausewitz was mainly concerned to understand and analyse the nature of war in its full political and social context. He was much less didactic than Jomini in his approach, notably as regards his lack of belief in the existence of permanent principles of war. His uncompleted *magnum opus*, *On War*, was much more difficult to understand than Jomini's publications, and contained profound problems of interpretation which continue to engage scholars. What most impressed studious officers in the later nineteenth century was Clausewitz's admiration for Napoleon's style of 'total war', his stress on the prime importance of the great battle as the optimum method of deciding operations, and his concern with leadership, will, and high morale as means of countering the friction, inertia, and uncertainty that characterized warfare. Clausewitz displayed more interest than Jomini in exploring and clarifying the political dimension of war. Indeed, in the last three years of his life (1827–30) he gave increased emphasis to the idea that war is an instrument of policy and has no rationale or logic of its own. Ideally, at the higher level of strategy, political considerations should inform and control the military instrument throughout hostilities.

This critical idea had obvious implications for civil–military relations whose problems both Clausewitz and Jomini had both experienced at first hand. But, in part perhaps because their active careers had been in an era when monarchs, princes, and, outstandingly, Napoleon himself had served as both military and political leaders, they left the institutional arrangements for the higher direction of war rather vague.

Thus my title 'The Pursuit of Victory', with the implicit addition of the phrase 'through decisive battle', had its origin in the Napoleonic example inculcated as orthodox military doctrine and disseminated to successive generations of European and American officers through the works of Jomini, Clausewitz, and many less well-known critics and publicists. It is no exaggeration to say that, as regards large-scale wars between major powers, it was the dominating strategic concept throughout the nineteenth century. Although from a modern standpoint many 'limited' wars were fought (for example, the Crimean War in 1854–6, the campaign in northern Italy in 1859, and the Danish war in 1864), there was no coherent doctrine to govern their conduct.

As a title 'The Pursuit of Victory' may also suggest an ironic addition or qualification such as 'a mirage' or 'will-o-the-wisp', and this for two reasons. First, in the purely military sphere, it would become increasingly difficult in the later nineteenth century for commanders to win victories which were 'decisive' in the sense that they annihilated the enemy's main army or battle fleet to the extent of making further organized resistance impossible. The revolution in fire-power, the rapid spread of railways and the telegraph, and perhaps most significant of all, the ability of industrialized nation states to raise and maintain huge conscript armies and echelons of reserves—all these factors suggested that wars would be decided more by attrition than by decisive battles.

Secondly, the notion of 'decisiveness' implies political direction and control. The statesman must be in a position to end hostilities at an opportune moment; persuade the beaten enemy to accept the verdict of battle; and reach a settlement which is not only acceptable to the warring parties, but also to other interested parties who may otherwise interfere to overthrow the settlement and perhaps even combine against the victor.

The Prussian/German victories over Austria in 1866 and France in 1870-1 constituted remarkable examples of decisive warfare which seemed to suggest that the Napoleonic ideal could still be realized under changed conditions. As the discussion in Chapter 4 shows, the military facet of the victories owed a great deal to Prussia's superior organization, training, weaponry, and command-staff system, superbly directed by the elder von Moltke as chief of the general staff. But these victories would also have taken a very different direction without Bismarck's successful challenge to the military doctrine of professional dominance in wartime: particularly his precisely formulated and limited aims; and not least his ability to persuade the defeated governments and interested non-belligerents to accept the verdict of battle. These were also 'decisive' victories in that the verdict of 1866 was never reversed and that of 1870–1 only in part as a result of the wider conflict of 1914–18.

In a longer perspective the nature, duration, and outcome of the American Civil War (1861–5) is generally agreed by historians to be a more accurate pointer to the character of modern war.[3] At the outset the South possessed superior generals, notably the combination of Robert E. Lee and Thomas 'Stonewall' Jackson in Virginia, and superior all-round combat skills, which resulted in a series of brilliant victories, including First and Second Bull Run, Fredericksburg, and Chancellorsville. But, gradually, Northern leadership improved and the advantage in terms of population, railways, warships, industrial, and financial strength began to tell. Fundamentally, however, the victory of the North was due not so much to the operational skills of its generals but more to its capacity to mobilize such industrial and manpower superiority as to render the South's operational advantages almost irrelevant. In sum, 'the *logistical* dimension of strategy proved more significant than the operational'.[4] Since the numerically and economically weaker South would not give in, its territories were divided, its ports blockaded, and its resistance gradually ground down by a combination of battles and more general attrition of manpower, armaments, and supplies. Grant was not a particularly skilful tactician, but displayed a remarkably 'modern' grasp of the relations between strategy and policy in an unlimited conflict which did eventually result in a clear victory for the North.[5] The main lesson, which European critics were understandably reluctant to draw, was that where mass armies clashed over vital issues of national sovereignty and culture, the war would rarely be decided by a decisive battle or battles. Rather battles would constitute an important part of the attritional struggle to exhaust the weaker side over the long haul, measured in years rather than in months. Whether such protracted and extremely costly conflicts really served the ends of 'policy' was a difficult issue for statesmen and, ultimately in democracies, public opinion to decide.

To his credit, the elder Moltke became increasingly doubtful after 1871 as to whether his triumphs could be repeated in a future two-front war. His successor but one, Count Alfred von Schlieffen, decided that Germany must strive for a quick and decisive victory on one (the Western) front in a two-front war, but there is more than a hint of anxiety and even desperation about German and European military planning in general before 1914. The impending great war *had* to be won quickly and decisively by great battles on the frontiers because the consequences of failure were too horrendous to contemplate.

It is tempting, indeed almost unavoidable, to criticize pre-war generals in all the belligerent nations of the First World War for their apparent obsession with 'the offensive' (in strategy and tactics) as a means to win a 'decisive victory'. But it needs to be remembered that, at the operational level, professional soldiers have a duty, except on rare occasions when a purely defensive strategy is decreed, to plan and prepare for victory. The Japanese defeat of Russia on land and at sea in 1904–5 provided a timely, impressive example of how the offensive spirit could prevail over a supposedly more powerful opponent, whose officers and soldiers displayed a less fanatical will to win.

It is rather in the grey area where military and political responsibilities overlap and intermingle that the generals (especially the pre-1914 generation) may be criticized in so far as they usurped the statesman's role by urging war in peacetime; insisted on autonomy in the professional sphere in wartime without regard to political implications; or insisted on military goals which proved unattainable. German and Austrian generals, but by no means they alone, were notably guilty of these professional shortcomings.

The quest for decisive victory through battle proved illusory in 1914 for several reasons including the vast numbers involved, the geographical factor (too much space in Eastern Europe not enough in the West), and the will and capacity of the governments and peoples to keep the war going. But the most important consideration, in relation to the theme of this book, is that the governments of the belligerent nations had all, more or less willingly, abandoned responsibility for strategic decision-making to their commanders before or at the onset of war. Forceful and combative civilian leaders, such as Lloyd George and Clemenceau, eventually emerged, but even they found it difficult to impose alternative strategies. By 1916 the military commanders had set the agenda: an attritional struggle for victory and ambitious war aims which did not allow for a negotiated peace. The war witnessed many successful operations, including Tannenberg, the Brusilov offensive, the British surrender to the Turks at Kut, the German overrunning of Romania, Caporetto, Cambrai, the German March 1918 offensive, and the August battle of Amiens, but none proved 'decisive' and some, such as Ludendorff's March offensive, were strategically disastrous. This is not to argue that these and the many other set-piece battles were not important in affecting the final results in the different theatres; merely that in such an enormous global conflict between alliances, no battle could be decisive once the original offensives had failed. The outcome could only be decided, given the maintenance of national morale and the will to continue—which even the Habsburg Empire displayed until the final weeks—by the 'big battalions' measured not only in manpower but also in industrial, financial, and maritime power.

Although field commanders such as Foch and Haig very properly continued the quest for victory, the Napoleonic ideal of the 'knock-out blow' was widely discredited by the attritional character of the First World War. The victories eventually achieved belied the hopes and expectations of 1914. Readers will notice that the emphasis of this study switches from the operational level to that of grand strategy and politics. This is done to press home the point that, despite the appalling casualties and destruction, the

First World War did conclude with clear winners and losers. Germany obtained a decisive victory over Russia and imposed draconian peace terms in the Treaty of Brest-Litovsk in February 1918. In the following months Britain and France faced the dreadful possibility that Germany would also secure a victory in the West, with consequences that would threaten their future security. But even a stalemate in the West would leave Germany dominant over a vast area of Eastern Europe and what shortly became the southern states of the Soviet Union. The remarkable Allied victory on the Western Front, muffled by the fact that Germany signed an armistice just before her homeland was invaded, not only thwarted the latter's territorial ambitions in the West (and returned Alsace-Lorraine to France), but also in effect negated the terms of Brest-Litovsk.

This was essentially a negative victory for Britain, France, and Belgium— they had avoided defeat and seemed to be in a position to impose severe terms which would prevent a resurgence of German militarism. But the United States' intervention in the war in April 1917 injected more idealistic goals, which greatly complicated the issues of war aims and peacemaking. That such hopes of lasting peace, enforced by international pressure and sanctions soon proved delusory, should not be allowed to obscure contemporary relief and rejoicing that such a terrible struggle had ended in victory. The bitterness and disenchantment, and even feelings of betrayal, at the meagre fruits of victory in domestic and international politics, should not be read back into the military effort itself. Too much had been promised, especially in extravagant propaganda and rhetoric, which simply could not be delivered in the conditions prevailing after 1918.

One consequence of the speedy post-war revulsion against the conduct of operations was that the reputation of the military leaders suffered a decline in popular esteem. Indeed, few of the generals themselves—especially in the victorious democracies—retained much confidence in their ability to win another great war.[6] It soon became apparent that the experience of 1914–18 constituted a turning-point in popular, professional, and political attitudes to war. Henceforth it became very difficult to think coolly about resorting to war 'as an instrument of state policy'. Military aggression to alter frontiers was prohibited by the League of Nations' Charter, and in the late 1920s the resort to war was formally renounced by signatories of the Kellogg–Briand Pact. Jomini's long predominance in the shaping of European and American military minds came to an end. Clausewitz's reputation also suffered an eclipse among the victors, though he was still respected in Germany where significant political and military factions did not accept the military result of 1918 as final.

The liberal belief that 1914–18 had demonstrated the futility of war and that no such large-scale catastrophe would be perpetrated again among 'civilized nations' took a deep hold in certain countries, only to experience a terrible shock in the late 1930s. The *Wehrmacht*'s blitzkrieg campaigns between 1939 and 1941 in Poland, Denmark, the Low Countries, Norway, France, and the Balkans, and Japan's remarkable Pacific conquests between December 1941 and June 1942, demonstrated that militarily decisive victories on a Napoleonic scale were still possible under modern technological and political conditions. Mechanized ground forces, closely supported by tactical air power, demonstrated that mobility and decisiveness, in the military sphere, had been restored to warfare. Whether such relentless and ferocious war-making was compatible with rational political control may be doubted; German and Japanese professional excellence was not accompanied by the political moderation and wisdom that might have converted their conquests into lasting gains.[7]

For three years their opponents suffered a series of humiliating defeats with very little ability to retaliate, other than by strategic bombing, but gradually the tide turned and Germany, Japan, and Italy were forced to contemplate defeat. Battle victories such as El Alamein, Guadalcanal, and Imphal were important for military and civilian morale, but it was clear from an early stage that even a huge victory on the scale of Stalingrad or Kursk would not be decisive in itself. In addition to the war-long naval and air campaigns, the Axis Powers were exhausted by essentially attritional campaigns in the Middle East, Italy, Burma, the Pacific, and, far surpassing all the rest, the Eastern Front.

To cover these operations in any detail would be to risk writing an extensive history of the war, so I have preferred to concentrate on more controversial issues on the level of grand strategy. Half a century after the end of the war it is easy to argue that, since the Allies were bound to win from, say, mid-1942 onwards, they should have adopted more limited and humane strategies which would have caused less destruction and fewer losses, thereby also reducing the legacy of bitterness.[8] In my opinion such an approach is unhistorical, not only in underestimating the emotions prevalent at the time but also, more seriously, in underrating the real difficulties of maintaining the anti-Axis alliance until unambiguous, decisive victories could be won in both Europe and the Pacific. Rather than place the main responsibility on the Allies for the policy of 'unconditional surrender' or for the strategy of conventional and, ultimately, atomic bombing, the German and Japanese leaders deserve criticism for prolonging the war needlessly at the expense of their own populations. The dreadful experiences of the years 1943–5, for all the belligerents, illustrates in extreme form the problem already evident in the nineteenth century and the First World War when powers which have deliberately resorted to war as an instrument of policy prove unwilling to accept that they have been beaten and must admit defeat.

The notion that military victory would retain its relevance in the nuclear era seemed implausible to some military theorists after 1945, but it soon became clear that, although nuclear powers were most unlikely to fight each other, they would not be able to deter conventional wars in many cases where only one or neither side possessed nuclear weapons. In a wide variety of conflicts fought for independence from imperial rule, for religious or territorial reasons, or simply to preserve the territorial integrity of existing states, winning—or at least not losing—seemed as important as ever. It became, if possible, even less permissible than before 1939 for governments to talk openly of using war 'as an instrument of policy', but such was clearly the case in, for example, the Korean War, the Suez crisis, the Argentinian attempt to seize the 'Malvinas' islands, and Saddam Hussein's annexation of Kuwait.

What has become equally clear since 1945 is that resort to war has ceased to be a valid option for some states as members of alliances dominated by superpowers, or simply because war can serve no practical purpose in furthering their interests. It has also become harder for smaller powers to win complete military victories unless acting under a United Nations mandate or with the informal protection of one of the superpowers.

Since 1948, however, Israel has been a notable exception in winning a series of military victories by employing the standard methods of modern warfare including: strategic surprise (except in 1973), superior air power, extremely efficient military organization, leadership, high morale, and rapid manœuvres in the field making good use of mechanized forces. Moreover, until her incursions into Lebanon from the early 1980s, Israel enjoyed widespread international sympathy as the underdog and was consequently able to enhance her territorial security. In recent years, however, Israeli governments have been under domestic and international pressure to negotiate over the occupied territories such as the Gaza Strip, the West Bank, and the Golan Heights. But, as a consistent winner, she is in a strong position to safeguard what are considered to be vital interests against states and organizations which have been unambiguously determined to destroy her.

Critics and commentators who have predicted the end of conventional warfare—perhaps more in hope than expectation—have quickly been proved wrong by large-scale conflicts in the Middle East (Iran versus Iraq), the Falklands, and the Gulf, not to enumerate the long list of civil wars, armed insurrections, and a variety of 'low-intensity conflicts' now raging in the Balkans and elsewhere. While a few areas are likely to remain free from such wars, this still leaves all-too-many areas in Africa, the Middle East, Asia, and the former Soviet Union where conflicts are impending or already endemic. Such conflicts are very different in form and style from those with which this study started, but many are still essentially Clausewitzian in having political objectives which, the antagonists believe, can only be realized through military victory.

Frederick the Great and the Era of Limited War

If we are to appreciate the tremendous impact made by the French Revolution and Napoleon on both the theory and practice of war it is essential
to describe the character and ethos of war in the preceding era. These can
most conveniently be viewed through the exploits and experiences of the
outstanding soldier-statesman, Frederick the Great of Prussia (1712–86).
Historians agree that there was a great deal of continuity between, say, the
1750s and 1790s in the steady improvement of physical conditions, notably
of roads; of weapons such as field artillery, and, perhaps most important of
all, of military organization.[1] Continuity is also evident in the publications of
military theorists, some of whom, by standing back from the esoteric details
of tactics, siegecraft, and logistics, partly discerned that these developments
collectively gave armies a greater potential for achieving decisive military
and political results. Indeed, the most perceptive of the theorists, the Comte
de Guibert, seemed in retrospect to have predicted the main characteristics
of revolutionary and Napoleonic warfare with almost uncanny accuracy.[2]
Guibert's most original insight, from whose implications, however, he drew
back in aristocratic alarm, was that the key element needed to revolutionize
purely military developments was the harnessing of popular allegiance and
energy to the state. This would transform what had previously been limited
dynastic conflicts between rulers and their mercenary armies into unlimited
ones between nation-states and their peoples in arms.[3]

While Guibert and other later eighteenth-century theorists could only
imagine and predict these far-reaching changes in the character of war,
Clausewitz was the first and profoundest analyst of the transition from
limited to 'total' war after it had been implemented by Napoleon.
Clausewitz perceived that political and economic conditions under the
ancien régime were inimical to ambitious military objectives.

Even a royal commander had to use his army with a minimum of risk. If the army was pulverized, he could not raise another, and behind the army there was nothing. That enjoined the greatest prudence on all operations ... The conduct of war thus became a true game, in which the cards were dealt by time and by accident ... Even the most ambitious ruler had no greater aims than to gain a number of advantages that could be exploited at the peace conference.[4]

Moreover, a balance of power had developed between the larger European states which militated against drastic changes effected by victory in battle: 'Political relations ... had become so sensitive a nexus that no cannon could be fired in Europe without every government feeling its interest affected.' Finally, he noted that ruthless aims in warfare, including plundering and laying waste the enemy's land, and harsh treatment of non-combatants, were now contrary to the more civilized spirit of the age. The result had been, in the mid-eighteenth century, that armies, relying heavily on defensive tactics behind their fortresses and prepared positions, had formed a state within the state, in which violence gradually faded away. Battles tended to occur only in relation to such modest aims as seizing or defending a fortress. Anyone who fought a battle that was not strictly necessary, but simply out of innate desire for victory, was considered reckless.[5]

Europe rejoiced at this development, which was seen as a logical outcome of Enlightenment, but Clausewitz regarded this as a misconception because it depended on all states observing the same code of practice. As a young man (he was born in 1780 and joined the Prussian army at the tender age of 12), Clausewitz was deeply affected by the humiliating defeats inflicted on his own state by the French revolutionary armies and, above all, by Napoleon in the Jena campaign of 1806, because Prussia's leaders had failed to grasp that war was now being waged with a new dynamism and ferocity, and for far higher political stakes than in the previous era.

Although Clausewitz occasionally berated eighteenth-century commanders for their half-hearted and even pusillanimous conduct of operations, he generally excluded Frederick the Great from his strictures: it was the fault of his successors that the army had been allowed to atrophy. The King of Prussia had displayed wisdom in that, pursuing a major object with limited resources, he did not try to undertake anything beyond his strength, but always did *just enough* to get him what he wanted, that is to seize and then retain the Austrian province of Silesia. 'His whole conduct of war, therefore, shows an element of restrained strength, which was always in balance, never lacking in vigour, rising to remarkable heights in moments of crisis, but immediately afterward reverting to a state of calm oscillation, always ready to adjust to the smallest shift in the political situation.'[6]

There is general agreement that eighteenth-century warfare was limited in comparison with the religious wars that preceeded it and the revolutionary and nationalist wars that followed. Thus a French soldier writing in 1792, only a year before the *levée en masse* opened a new era, remarked that: 'They are no longer nations that fight, nor even kings, but armies and men payed [*sic*] for fighting: it is a game, where they play for what is staked and not for all that they have in the world; in fine, wars which in old times were a madness, are at present only a folly.'[7] In an enlightened age which believed in both spiritual and material progress, it seemed not merely foolish and counter-productive, but also morally wrong for war to be waged as a ruthless, all-out struggle involving great physical damage and civilian casualties. In the operational sphere, wars were often long but seldom intense; and battles, fought at point-blank range, could cause horrendous casualties, but were often—for that very reason—avoided by the weaker side. Increasingly, in the middle decades of the century, operations focused on fortresses, magazines, and supply lines, producing a warfare of almost mathematical formalism in which skilful manœuvre was valued above the determination to fight. Rulers and commanders (sometimes the same person) came to accept protracted campaigns yielding only modest advantages.

However, the practice of war and humane, civilian values are inherently at odds, so the notion of moderation in eighteenth-century campaigns must not be pushed too far. In particular the fragile economic and financial base of most dynastic states could in some cases put a premium on short wars where ruthlessness, even for modest aims, might be necessary to force the enemy to negotiate. In the war in Piedmont (1742–8), for example, the French army's harshness in requisitioning supplies provoked guerrilla operations against their communications. Joseph II of Austria was devastated by his first experience of warfare in 1778: the destruction of fields and villages, the lamentation of the poor peasants, and the ruin of so many innocent people, impressed on him the real frightfulness of war.[8] Indeed, the historian must frequently remind himself or herself that, no matter how limited political aims may be, or the willingness of the belligerents in principle to observe humanitarian codes of behaviour, the consequences for combatants or civilians caught up in the action are always likely to be appalling. More soldiers died of disease than in combat and the primitive means available for treating the sick and wounded hardly bear thinking about.

Although historians have generally accepted the term 'limited war' as applicable to the decades before the outbreak of the French Revolution, there has been an unresolved debate about the fundamental cause of such limitations as there were. Was the general character of warfare determined by such broad considerations as the dynastic states system reinforced by the Enlightenment culture, philosophy, and ethos; or did the predominant influence lie rather in material conditions such as meagre economic and financial resources, poor communications, unwieldy armies which had to be kept united for fear of desertion, and a tactical system—itself conditioned by cumbersome weapons—which made it extremely difficult to convert battlefield superiority into a decisive victory?

Jean Colin, a French soldier and pioneering scholar of pre-Napoleonic military history who was killed in the First World War, asserted confidently that it was the means and methods which determined the character of war, not the limited aims of the rulers. He ridiculed the latter view as absurd. Why, he asked rhetorically, would such energetic leaders as Louis XIV or Frederick II have gone to the enormous expense of raising and arming troops for such modest gains had they known a way of crushing their enemies more promptly? By betraying the tacit pact attributed to them one of those sovereigns would have been able 'to achieve more power and glory in six months than Napoleon did in ten years'. No, in his opinion, based on considerable archival research, weapons determine the manner of fighting and hence tactics; this in turn determines the organization of armies and their ability to manoeuvre; and, ultimately, these factors shaped the character of the entire war. Colin derided, in particular, the view that great generals had ever allowed themselves to be dominated by logistical considerations. A modern scholar, Hew Strachan, has endorsed Colin's standpoint by writing 'If the effects of battle were limited, this was due to the practical constraints of the times and not to design', and 'The means were not limited if they were consonant with the objectives'.[9]

The contrary view, which probably holds the support of most students of the issue, is that although material, technical, and purely military innovations provided essential underpinning for the new style of warfare, it was the political revolution after 1789 which made the essential difference between the 'limited' war of Frederick's heyday and the 'total' war conducted by Napoleon.[10]

Clausewitz, as was noted earlier, took a historicist view that each era would manifest its own unique style of warfare, shaped by such variables as inter-state relations, the economic and manpower resources available, and the moral-legal-religious climate. But in the end, in the light of his experience and reflection, he gave emphatic priority to popular participation in the affairs of state as the main reason for the new style of warfare in his time. Suddenly, from 1793 onwards—'a force that beggared all imagination'—the people became a participant in war so that the full weight of the nation was thrown into the war effort. Nothing now in principle impeded the vigour with which war could be waged and, consequently, France's opponents were placed at a crippling disadvantage. Clausewitz of course knew that, even in the new conditions, war was sometimes conducted in a timid, indecisive way, but for him Napoleonic 'total war' closely approached its true character, its absolute perfection.

War, untrammeled by any conventional restraints, had broken loose in all its elemental fury. This was due to the peoples' new share in these great affairs of state; and their participation, in turn, resulted partly from the impact that the Revolution had on the internal conditions of every state and partly from the danger that France posed to everyone.[11]

Our view of the eighteenth century as an era of limited war must be modified by the fact that it was 'an age of battles' with, surprisingly, a higher ratio of casualties to the total of participants than in the Napoleonic or mid-nineteenth-century wars in Europe.[12] Frederick the Great usually fought with inferior numbers and often suffered higher casualties even when tactically successful. For example at the battle of Prague in May 1757 the defeated Austrians lost about 10,000 killed and wounded and more than 4,000 prisoners, whereas Frederick lost 11,740 killed and wounded and 1,560 prisoners.[13] For both armies these total losses amounted to approximately 21 per cent of their forces. At Kolin the following month, Frederick suffered defeat in a rare frontal assault losing 6,710 killed and wounded and 5,380 prisoners out of 33,000. His Austrian opponent, Marshal Daun, lost a similar number killed and wounded but only 1,500 prisoners out of 53,000. At his great victory over a French and German army at Rossbach (November 1757), Frederick lost only some 550 as against nearly 8,000 allied casualties, but this was exceptional. In perhaps his most famous victory—at Leuthen—in December 1757, Frederick could muster only about 43,000 troops against 72,000 Austrians. The latter's losses of 25,000 consisted mostly of prisoners, but the Prussians also lost 6,200, which brought their total casualties for the year to over 50,000. This was exorbitant for an impoverished state of only some four million. Finally, mention must be made of Prussia's two bloody encounters with the Russians. In the indecisive battle at Zorndorf in August 1758, Frederick lost about 14,000, killed, wounded, and missing, a casualty rate of over 37 per cent, but his opponents suffered 21,000 casualties—a staggering 50 per cent of their force. At Kunersdorf in August 1759 the Prussians were routed by an Austro-Russian force, suffering more than 20,000 casualties (to their opponents' 15,700) and also abandoned 278 guns. Several more bloody battles were to follow but we need only note here that by the end of 1760 the drain on Prussian military manpower had reduced the proportion of Frederick's own subjects to about one-third of the total in his armies and even with the recruitment of foreign mercenaries (and prisoners from defeated German states such as Saxony) his manpower shortage remained acute until the end of the war in 1763.

In sharp contrast to Napoleonic strategy, eighteenth-century campaigns rarely had the decisive battle as their culminating objective to which all previous planning and manœuvres were directed. Losses were difficult to replace, much time and effort was required to produce regiments trained to the exacting drill standards of the day, and armies were extremely expensive to maintain and supply in the field. Manœuvre could therefore all too easily become an end in itself as an alternative to battle, particularly as it was relatively easy for the weaker or more cautious enemy to avoid being cornered. Even with two armies face to face it took a long time to deploy the troops from marching columns into battle lines, so if one side chose to retreat at this late stage no complete engagement could occur.[14]

Frederick was among the most vigorous and determined commanders of his day, whose voluminous writings can be quarried to provide contradictory opinions on many issues, including the place of battle in his war plans. 'Battles decide the fate of a nation', he wrote. 'In war it is absolutely necessary to come to decisive action either to get out of the distress of war or to place the enemy in that position.' But he also advised caution, remarking that 'a general of an army will never give battle if it does not serve some important purpose'.[15] He was seldom in a position to risk full-scale showdown battles between the main forces because he frequently had to split up his army to meet multiple threats, and in any case felt that the outcome depended too much on chance. Consequently he rarely sought to destroy the enemy's main army, but rather aimed to force him to retreat. 'To win a battle', he wrote, 'means to compel your opponent to yield you his position.' Thus, in contrast to the short, sharp campaigns he had recommended in 1746, his practice tended increasingly towards complex manœuvre for the acquisition of modest gains. He also drew from his own unfortunate experience in Bohemia in 1757 the lesson that an army could not operate successfully far beyond its own frontiers. Compared with Napoleon's daring and reckless forays to the far corners of Europe, and beyond into the Mediterranean, Frederick's numerous campaigns with their remarkable marches and counter-marches took place in a surprisingly compact area in central Europe.[16]

It must be stressed that, though typical of the military orthodoxy of his age in his preference for wars of manœuvre and position over the quest for decisive battle, Frederick always scorned passivity or inertia, whether on the offensive or defensive. Even after the exhausting and personally desolating experience of the Seven Years War, he was still prepared in principle to repeat the bold pre-emptive strike of 1740 by invading Saxony or Bohemia.[17] He knew only too well that for a scattered and relatively weak state such as Prussia a purely defensive stance would invite military defeat and political obliteration, such as would befall Poland in the 1790s.

So far in this chapter we have tried to establish that, in comparison with the 'total war' practised by Napoleon, eighteenth-century warfare was generally limited both in terms of political objectives and of the armies' ability to deliver a knock-out blow in battle, followed by a relentless pursuit, in order to end a war in one dramatic stroke. Clausewitz accepted that in the pre-Napoleonic era it was difficult to win such a clear-cut victory, but he was nevertheless critical of generals who refused to take risks because he believed that 'only great victories have paved the way for great results; certainly for the attacking side, and to some degree also for the defence'. He regarded battle as a decisive factor in the outcome of a war or campaign, but not necessarily as the only one. Displaying remarkable historical objectivity he added: 'Campaigns whose outcome have been determined by a single battle have become fairly common only in recent times, and those cases in which they have settled an entire war are very rare exceptions.'[18]

In his recent study *The Age of Battles*, Russell Weigley has taken this sceptical approach to an unconvincing extreme by arguing that even in the period he has designated in his title—between Breitenfeld (1631) and Waterloo (1815)—war lacked the only virtue claimed for it, namely, the power of decision. He describes the period covered as one of prolonged, indecisive wars in which the toll in lives, not to mention the costs in material resources, rose grotesquely out of proportion to anything their authors could hope to gain from them. Warfare in this period, he allows, was indecisive in a different form from the First World War or Vietnam, but 'recalcitrant, intractable indecision nevertheless persisted'. The best opportunity for wars to be decisive at a reasonable cost was when the outcome could be resolved by one battle in an afternoon. 'If wars remained incapable of producing decisions at costs proportionate to their objects even then [i.e. between 1631 and 1815], consequently *the whole history of war must be regarded as a history of almost unbroken futility. So it has been.*'[19] This verdict is, to say the least, a surprising one for a distinguished military historian, and especially one who then devotes more than 500 pages of very detailed battle description to an activity which he regards as futile.

The main objection to Weigley's thesis is that his criteria for decisiveness are too stringent. As Clausewitz pointed out, it has rarely been possible to destroy the enemy's forces so completely in a single day's fighting that its government has felt compelled to sue for peace. But the history of warfare is replete with cases where a series of successful battles, complemented by other forms of pressure such as blockade, propaganda, and diplomacy, have eventually resulted in undisputed victory for one side and defeat for the other. Weigley is, of course, well aware of this fact, but he denies its validity by another criterion which is only implicit in his study; namely, that only an outcome of victory in battle which results in the permanent and final resolution of the sources of political conflict would merit the term 'decisive'. Thus, as a perceptive reviewer has pointed out,[20] even Marlborough's outstanding victory at Blenheim must be regarded as indecisive because it (and his other triumphs) failed to break France as a great power. This, it may be suggested, is to set impossible standards for the enduring success of particular actions not only as regards military history but for most other forms of human endeavour.

In the real world of historical experience, Weigley would surely concede that such battles as Blenheim, Plassey, Quebec, Austerlitz, Jena, Trafalgar,

Borodino, and Waterloo were decisive in specific ways which were evident at the time and have satisfied historians. Moreover, it is extraordinary that an American historian should exclude two of the most decisive battles of the eighteenth century, namely Saratoga and Yorktown. His underlying polemical thesis seems to be rather that most conflicts are protracted and consequently resolve disputes or advance policies at excessive cost, if at all. This is an entirely respectable standpoint for a moralist, but surely not for a historian who has to take the follies and imperfections of men and governments as his stock in trade.

Although Weigley's pessimistic, even despairing, view of the efficacy of war as a political instrument pervades his text in that, for example, he concentrates heavily on a narrative of military operations at the expense of the political context in which they occurred and were terminated, he none the less reveals that even long, intermittent, and seemingly indecisive wars (in the sense that there was no culminating battle), could still have significant, positive results. Such campaigns served to frustrate Louis XIV's and Napoleon's attempts to impose French hegemony on Europe, to cause Britain to recognize American independence, and to determine that Britain rather than France would enjoy colonial control over Canada and India. Most interestingly for the present chapter, Weigley—at variance with his own thesis—demonstrates that Frederick the Great skilfully and successfully employed war as an instrument of state policy throughout his long reign.[21] Frederick ruthlessly invaded the valuable Habsburg province of Silesia in 1740 without a declaration of war and, thanks largely to subordinates, won the battles of Mollwitz (1741) and Chotusitz (1742) to gain an alliance with France and retain his prize. In 1744 Frederick broke the Treaty of Breslau by invading Bohemia, where he occupied Prague but was then hard-pressed by superior forces and was fortunate to gain a tactical victory at Hohenfriedberg. After further indecisive engagements Prussian possession of Silesia was formalized by the Treaty of Dresden.

In 1756, at the outset of what was to become the Seven Years War, Prussia was confronted by apparently impossible odds in the form of a triple alliance between Austria, France, and Russia. The last-named, in particular, sought the unlimited goal of the elmination of Prussia as a major power. The war was consequently 'total' for Frederick in that the very survival of Prussia was at stake, yet 'limited' in the senses that his forces must be dispersed to meet this threefold threat and that he could never aspire to be strong

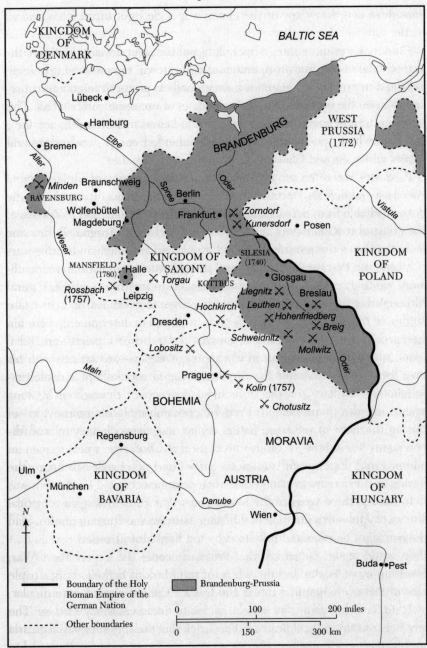

KINGDOM OF DENMARK

BALTIC SEA

Lübeck

Hamburg

WEST PRUSSIA (1772)

Bremen

Elbe

Aller

BRANDENBURG

× Minden
RAVENSBURG

Braunschweig

Spree

Berlin

Oder

Wolfenbüttel
Magdeburg

Weser

Frankfurt

× Zorndorf
× Kunersdorf

Vistula

Posen

MANSFELD (1780)

Halle

KINGDOM OF SAXONY

SILESIA (1740)

× Glosgau

KINGDOM OF POLAND

Rossbach (1757)

× *Torgau*

KOTTBUS

Leipzig

Liegnitz ×

Breslau ×

Hochkirch

Leuthen ×
× *Hohenfriedberg*

Dresden

× *Breig*

Schweidnitz ×
× *Mollwitz*

Lobositz ×

Oder

Main

Prague × ×
× *Kolin* (1757)

BOHEMIA

× *Chotusitz*

Regensburg

MORAVIA

Ulm

München

KINGDOM OF BAVARIA

AUSTRIA

KINGDOM OF HUNGARY

Danube

Wien

N

Buda ● ● Pest

—— Boundary of the Holy
Roman Empire of the
German Nation

Brandenburg-Prussia

- - - Other boundaries

0		100		200 miles
0	150		300 km	

Map 1. Central Europe during the campaigns of Frederick the Great

enough to overthrow any of the enemy states outright through success in battle.

Frederick's grand strategy (including political aims) was therefore consistently defensive, but his operational strategy and tactics were offensive, designed to prevent his enemies from uniting, to protect Prussian territory, and to drive the opposing forces back whenever an opportunity offered for battle on advantageous terms. It was this judicious mixture of manœuvres, rapid marching, positional attrition in defended camps, and occasional battles which earned Clausewitz's tribute, quoted earlier.

Frederick was of course by no means consistently successful. He experienced several defeats, or drawn battles which were so costly as to be barely distinguishable from defeats, including the indecisive slaughter at Zorndorf, the disaster at Kunersdorf, and the Pyrrhic victory at Torgau. Berlin was more than once temporarily occupied by Russian and Austrian contingents.

After 1758 Frederick's strategy and style of operations became markedly more cautious and defensive, as he began consciously to play for time. Although his soldiers remained capable of heroic exertions, as in the late battles of Liegnitz and Torgau, neither in discipline nor training were his later armies suited for the rapid offensives of the early campaigns. The King found it increasingly difficult to win clear-cut victories, as his infantry lost their former impetus while the Austrians became more adept at exploiting terrain and artillery for the defensive. Consequently, much as it went against his own temperament, Frederick became more circumspect in assessing the risks of full-scale battle. As his resources diminished and his opponents learnt how to counter his tactical methods, the war became less one of rapid marches and set-piece battles and more one of minor skirmishes and inactivity in strong defensive positions.

In the last three years of the Seven Years War Frederick's gravest problem was to prevent a junction of the main Austrian and Russian armies. Had the two allies co-operated fully to deliver a final blow it seems certain that they could have destroyed the Prussian forces as a prelude to the overrunning of Berlin and the whole of the Mark of Brandenburg. In the face of this overwhelming threat Frederick could only manœuvre desperately to save his army, by means of feints, night marches, and sudden advances. At this critical juncture Frederick's formidable reputation gained in his earlier campaigns played a vital part in saving Prussia from extinction.[22]

In 1762, with his army approaching ruin and his appearance 'resembling nothing so much as a demented scarecrow', Frederick was saved from almost certain defeat by the death of Tsarina Elizabeth and her succession by her sickly nephew Peter III, who so hero-worshipped the King of Prussia that he withdrew from the alliance, changed sides, and contributed a contingent of 18,000 Russians to the former enemy.[23]

This famous episode, which was to give Hitler the delusory hope of recovery in 1945 through a rift among his enemies following Roosevelt's death, suggests that the economically weak and geographically dispersed state of Prussia could not have survived for seven years against a genuinely united alliance of three major powers. Nevertheless, Frederick's unceasing marches and occasional battles played a crucial part in weakening his opponents' forces and keeping them at bay. Let us look briefly at the effects of his two outstanding victories of Rossbach (15 November 1757) and Leuthen (5 December 1757).

Weigley allows that the former victory over Imperial and French forces brought lasting benefits to Prussia, in that France for all practical purposes withdrew from the alliance and did not again interfere effectively on German territory until 1806. This first great victory over French arms was also a powerful morale booster for Prussia. The victory at Leuthen—perhaps the greatest of the century since it was won against a resolute and confident enemy—was diplomatically important in that it strengthened the existing British admiration for Frederick as the gallant underdog and caused the government to reach an agreement with Prussia in April 1758 promising the latter an annual subsidy of £670,000 and a mutual undertaking not to conclude a separate peace—an undertaking which Britain cynically disregarded in 1762 in making peace with France.

Frederick eventually achieved his limited political goals not as a result of a decisive battlefield victory but because Russia and France lost interest in the expensive game of destroying Prussia, and Austria was too weak to do so unaided. The Austro-Prussian Treaty of Hubertusburg in February 1763 effectively restored the *status quo ante bellum* thereby confirming Prussia's retention of Silesia and recognition of her status as a great power.[24]

Frederick's cold, pragmatic approach to war and strategy might seem so commonsensical as to raise no historical issues. But in reality his legacy was to play an extremely controversial role in Prussian, and later German, military theory. In the 1870s the distinguished military historian, Hans

Delbrück, provoked a long and intense debate by developing Clausewitz's late and unrevised ideas about the existence of two types of war 'limited' and 'total'. Delbrück deduced that there were also two distinct types of strategy: either annihilation or exhaustion.[25] The sole objective of the former was the decisive battle, whereas the latter offered the alternatives of battle and manœuvre—battle being merely one of several equally valid means of attaining the political objective, and not necessarily more important than the occupation of territory, commerce destruction, and blockade. Delbrück insisted that the strategy of exhaustion was not an easy option, rather that in certain historical contexts it was the *only* realistic mode of warfare available.

What really enraged Delbrück's critics, including Colmar von der Goltz, Theodor von Bernhardi, and other officers of the German general staff, was that he had numbered Frederick among the exponents of the strategy of exhaustion, rather than with Napoleon and other exponents of annihilation. This contravened general staff orthodoxy that the strategy of annihilation was the only correct system, as allegedly formulated by Clausewitz, and that Frederick was the precursor, if not the model, of Napoleon. Moltke's general staff understandably wished to find their inspiration in the King of Prussia. Delbrück's response, to the effect that this dogmatic attitude did Frederick a great disservice and, further, that if judged as an exponent of the annihilation strategy he would emerge as a third-rate general, only served to add fuel to the flames since he was now accused additionally of maligning a national hero.

Delbrück delighted in polemics so the controversy rumbled on inconclusively for twenty years in German military journals.[26] Its significance for the later nineteenth century, which will be explored more fully in Chapter 5, was that it revealed the general staff's extremely dangerous tendency to divorce a strategic system from its political context. In this respect, it may be suggested, Delbrück's error was not so much the posing of strategic options, as to allow even the possibility of a strategy of annihilation in a Europe of advanced industrial states capable of fielding conscript armies numbered in millions. True, Delbrück offered a theoretical solution to Germany's problem of a two-front war, but even so Frederick's Fabian strategy would have provided a less risky alternative.

Frederick's own comments can be quoted in support of either form of strategy, but an authority on his writings[27] concludes that he was not doctrinaire or wedded to any particular system. His instincts and preferences were for offensive action in strategy as well as tactics, but he quickly realized that 'in war, as in everything else, a man does what he can and seldom what he desires'. Such fundamental considerations as Prussia's inferiority in available manpower, and inability to concentrate exclusively on the defeat of one opponent, caused him to resort to a strategy of manœuvre calculated to exhaust rather than defeat his enemies.

It would be wrong to assume, however, that even had his resources permitted, Frederick would necessarily have opted for the Napoleonic strategy of overthrow. Unlike Napoleon, after his seizure of Silesia in 1740 Frederick waged no further war of conquest. Throughout the Seven Years War his aim remained 'to bring Silesia into the safe harbour of a well-guaranteed peace'. He made other references to possible conquests in Poland, Saxony, and even Western Europe, but mainly with a view to gaining assets for trading in the eventual peace negotiations. His later writings, including his *Reflections on Projects of Campaign* (1775) show him to be very much a man of his age. For example, even in imagined campaigns against Austria and France where he assumed a Prussian superiority of numbers, he only argued that by deep advances to threaten either capital he would thereby force his opponents to negotiate.

Frederick's last campaign against the Austrians in the War of the Bavarian Succession in 1778 was still as aggressive and far-reaching as ever in conception. But by then his health was deteriorating and his army had lost its edge, particularly in mobility. The Austrians remained behind their formidable defences, frustrating Frederick's attempts to set up a decisive battle. Consequently, the result was inconclusive. Frederick called it an 'insipid' campaign and his troops referred to it as a 'potato war'.[28]

Frederick's practice and reflective publications alike testify to the fact that the idea of total victory did not accord with his belief in the balance of power, which, he felt, would prevent Prussia from making any further substantial gains from inherently more powerful rivals. It had taken the utmost effort over twenty-three years to cling on to Silesia. Above all, as the nineteenth-century German general-staff theorists were to neglect to their cost, Frederick the general always subordinated his strategy to the policies of Frederick the king. It was not so much the battles won by Frederick that made him a great general, vital though the outstanding examples were in gaining him breathing space, but rather his admirably balanced political judgement and the conformity of his strategy with political reality.

Frederick himself brought out this point by implication in his criticism of the otherwise heroic military exploits of Charles XII of Sweden whom he greatly admired. Had Charles possessed moderation equal to his courage and set limits to his triumphs he could have secured an honourable peace with the Tsar. 'Unfortunately the passions of that man were subject to no modification. He wished to carry everything by force and haughtiness, and despotically to triumph even over despots.'[29] These words might equally have been applied to Napoleon whose heroic quest for total victory would exert such a profound influence on military theory for a century after the Emperor's final defeat.

It was Frederick the Great's achievement to develop the art of war to the maximum limits within the framework of existing social and political institutions. His successes on the battlefield were based on relentless attention to drill and discipline, aimed at producing an army which could manœuvre with machine-like precision; and on his celebrated oblique order tactics which, at best, would unhinge the enemy firing-line and lay it open to a devastating cavalry charge. He was seldom able to clinch his victories on the field by a vigorous pursuit; indeed only that after Leuthen approached the ideal set forth in his writings. Frederick was not by temperament an innovator, and in any case flexibility was severely restricted by the problem of desertion in a motley army composed of unwilling conscripts supplemented by mercenaries and prisoners of war.[30] Even at the time of acutest manpower shortage during the Seven Years War Frederick would not consider a *levée en masse* (general conscription). His demeanour towards his troops was cold and formal and he made no pretence to charismatic leadership. He developed no divisional or corps system and his senior officers were not encouraged to use their initiative. The contrasts between Frederick's style of leadership, military organization, and tactics and those of the French revolutionary armies and Napoleon are therefore striking.

Frederick had disciplined himself from an early age to play the role of military commander, displaying the virtues of resolution, perseverance, and a façade of stoic calm in the face of adversity. But his character was emotional, artistic, and sensitive; he lacked sang-froid in battle. His often-expressed longing to retire to the quiet life of the mind at Sans Souci was genuine. He was prone to melancholy and depression: lost battles plunged him into despair and contemplation of suicide, but a strong sense of duty and responsibility to the state drove him on. Despite these personal torments he stubbornly refused to make any concessions to end the Seven Years War.

When the war did eventually end in triumph for Prussia, with Silesia retained and French power weakened, Frederick returned to Potsdam prematurely aged and taking little pleasure in his success. Though aged only 51 in 1763 he is described as: 'Gray, desiccated, arthritic, shabby in a worn uniform, his face now deeply lined . . . a man despising men, who spread coldness and fear around him.' 'My last days are poisoned', he wrote, 'and the evening of my life is as horrible as its morning.'[31]

As Clausewitz appreciated, Frederick had exploited to the full his advantages in the combined role of king and commander-in-chief. He used war coolly and cynically to expand and then defend his state against more powerful rivals. Unlike Napoleon he made no claim to any ideological mission, whether in terms of nationalism or the ideals later thrown up by the French Revolution. He was engaged in a ruthless pursuit of power in which there was no room for sentiment: Frederick broke treaties when it suited him and was in turn abandoned by his ally, Britain, at a critical point in the Seven Years War.

In most respects there was no sharp divide between the era broadly characterized as one of 'limited war' and the revolutionary 'total war' of the 1790s. In Prussia after 1763 military reform made only slow and halting progress, but in France—as one would expect from the previously dominant power whose recent efforts in land and sea warfare had been disappointing—there was a ferment of reformist literature accompanied by important practical innovations in artillery, drill and tactics, and communications. Most significant of all, France pioneered the way in the development of the divisional system which would transform military operations by enabling armies to manœuvre rapidly in all-arms formations capable of defending themselves and to converge on the battlefield, putting a less well articulated opponent at a serious disadvantage.[32] The divisional system, moreover, made it extremely difficult for the weaker army to slip away and avoid battle, thus facilitating the decisive encounter in which the loser would be not merely damaged, but destroyed. The essential ingredient still lacking in this potent mixture was supplied by the French Revolution; nationalist fervour, the mobilization of citizens in the 'nation in arms', and the resort to a rapid and ruthless style of war as a means of extending French power and political ideals throughout Europe.

Napoleon and the Decisive Battle

The battle of Valmy in 1792 marked the dawn of a new era in military history. Between that date and 1815 Europe witnessed almost uninterrupted warfare between France and virtually all the other powers of a scale and intensity not experienced for centuries. This transformation in the direction of 'total war' made the 'temperate and indecisive contests' of the era of the Enlightenment appear as absurd anachronisms.[1]

The French Revolution rapidly incorporated and gave added impetus to a series of reforms and innovations in military institutions and practice which were already well-advanced in the 1780s. By far the most important of these in its immediate impact was the Convention's adoption of universal conscription, implemented by War Minister Lazare Carnot's proclamation of a *levée en masse* in 1793. This rapidly produced a great increase in the size of the French army (732,000 by April 1794), which gave extra weight to an expansionist foreign policy and enabled French generals to fight more aggressively and more frequently. A significant corollary was that these huge citizen armies were obliged to break free from the civilized but constricting eighteenth-century practice of depending on depots and magazines for food and other supplies. Living off the country often meant in practice marauding or dying of hunger; but also signified a new tempo in war and a new determination to smash the enemy forces by ambitious manœuvres designed to achieve concentration on the battlefield—operations which were made feasible by the developing practice of organizing mass armies into divisions and corps.

For a few years the French revolutionary armies enjoyed immense advantages against the monarchist coalitions which opposed them. Nationalist fervour and revolutionary ideals were combined with innovative tactics, ambitious and politically acceptable commanders, and a modernized staff system which made possible the control of widely dispersed but mutually supporting divisions and corps. Consequently the French enjoyed some striking successes against politically brittle coalitions which failed to concentrate their available forces for the decisive battle, fought for limited aims, and lacked popular support.

The French armies soon overran the Austrian Netherlands and penetrated the Rhineland; and in 1796–7 the 26-year-old Napoleon Bonaparte carried out his first brilliantly successful independent operations in northern Italy. However, as Peter Paret has recently reminded us, these early successes must be kept in perspective. The revolutionary armies suffered almost as many defeats as they won victories. By the summer of 1799 Napoleon's Egyptian expedition had ended in fiasco, his Italian conquests had mostly been reversed, and the Austrians again controlled southern Germany. The French conduct of war was certainly superior to that of the *ancien régime*, 'but even with the experience of a dozen campaigns it was a qualified, not an absolute superiority'.[2]

Although Napoleon eventually emerged from a group of talented young commanders, including Jourdan, Hoche, and Moreau, any one of whom might have seized political control of France, Paret is surely persuasive in suggesting that, had he died young or been taken prisoner before attaining supreme power, France's subsequent history would have been profoundly different. France might have expanded to her 'natural frontiers' on the Rhine and including the Low Countries, but she would not have embarked on a quest for European hegemony. 'The Revolution and the transformation of war would still have left France the most powerful country in Europe, but a country integrated in the political community, rather than dominating and, indeed, almost abolishing it.'[3] By harnessing and exploiting the full military and political potential of the Revolution as developed by 1800, Napoleon was empowered to unleash a form of warfare of unprecedented boldness, geographical range, and intensity, so earning Clausewitz's accolade of 'the God of War'.

As Spenser Wilkinson, Jean Colin, and other historians established long ago,[4] Napoleon was fortunate to achieve high command just at the point when the numerous military reforms and civil developments of the 1780s had come to fruition and when the chaotic upheavals of the early revolutionary armies had largely been sorted out. Napoleon was not a tactical or technical innovator and it is fair to say that he mastered Europe with weapons available to other men before he rose to power.[5] For example, he was the beneficiary of a long and intense debate on infantry tactics which, after exposure to trial and error in the battles of the 1790s, had yielded a 'mixed' organization of light infantry (skirmishers), dense columns for marching and the attack, and linear formations. For a time this complex and flexible system was markedly superior to the obsolescent linear tactics of the *ancien régime* forces. As an artillery officer himself, Napoleon also exploited to the full the reforms of that arm carried out by Gribeauval, Du Teil, and others since the 1760s.[6]

But it was the articulation of the formerly unitary and unwieldy mass armies into divisions and corps, first proposed by Pierre de Bourcet and implemented by the Duc de Broglie in the 1760s, which proved to be Napoleon's most valuable inheritance. Autonomous divisions of all arms and with their own staffs, moving along separate roads, would in principle be mutually supporting but capable of independent action for brief periods. Thus were armies potentially transformed into instruments for rapid advance, flexible manœuvre, and decisive action when brought together on the battlefield. In the brief respite between 1801 and 1805 Napoleon reorganized all his land forces into army corps composed of two or three divisions each containing some 8,000 troops. Divisions were further divided into two brigades each with two regiments. Widely distributed throughout French-occupied Europe between campaigns, these corps and divisions could only be assembled, moved to the theatre of operations, and concentrated on the battlefield by immensely complicated calculations of time, space, and terrain and by meticulous staff work. In the hands of less able or more circumspect commanders such calculations proved to be a fatal handicap, resulting in detachments, dispersion, and defeat.[7] A large element of Napoleon's military genius lay in his superlative skill in implementing the corps and divisional system.

Napoleon was a supreme egotist, possessed of unbounded self-confidence, optimism, and calmness under stress. He seems to have accepted the heavy losses that resulted from his orders with equanimity. For the greater part of his adult life he was dedicated to the practice of war and utterly absorbed by it. He possessed to an unusual degree the dynamism and strength of character essential to overcome the friction and inertia inseparable from all military activity. In sharp contrast to the previous age he epitomized the traditional warrior in his determination to conquer or perish with glory. He was completely wrapped up in his own destiny. Robust health and youthful vigour were vital components in his early campaigns—

and indeed until 1812—enabling him to perform remarkable feats of endurance. Martin Van Creveld, for example, cites his spending ten days in a tent in subfreezing temperatures before Austerlitz; and covering 150 miles on horseback in forty-eight hours in Spain.[8] His capacity for continuous work, often going without sleep for several days, was prodigious. These qualities of intellect, physique, and a charismatic personality, allied to a mastery of professional skills and outstanding talents as a military leader, help us to understand Clausewitz's extravagant reference to Napoleon as 'the god of War' or Van Creveld's more measured description of him as 'the most competent human being who ever lived'.[9]

Napoleon's strategic outlook was ambitious and aggressive to an extent that denoted the beginning of a new era in warfare. His objective was invariably the destruction of the enemy's main field force as the speediest means of overthrowing his government, rather than the occupation of territory or the capture of the capital city. His strategic deployment was carefully planned to set the stage for the great and decisive battle. He fully grasped the error of France's enemies in the 1790s in mobilizing and deploying their forces piecemeal in successive waves and in wasting units in garrisons and detachments.[10] Hence his determination—later to be somewhat simplistically interpreted—to bring superior numbers to bear at the decisive point.

This climactic point was only reached, however, after wide-ranging and complex manœuvres, which were meticulously planned and reconnoitred. At the outset Napoleon's corps might be dispersed over the entire theatre of operations, perhaps 150 miles or more. When the enemy's main body was located the French corps would begin to converge, adopting a loosely quadrilateral formation. In optimum conditions, with about one day's march between the corps, the enemy would be engaged and fixed by the leading formation while those following fell upon the flank or rear. When facing a numerically stronger foe, Napoleon usually attempted to seize a central position, so dividing the enemy army, and while pinning down one part used his reserves to gain local superiority and smash the remainder. If, however, the enemy was of equal or inferior strength, Napoleon adopted an envelopment strategy against his flanks and rear. Ideally, though this was only completely achieved on four occasions, victory in the decisive battle was followed by a vigorous pursuit which completely destroyed the enemy's capacity to resist. Brilliantly successful manœuvres leading to victory in battle did not necessarily obviate heavy casualties, but until late in his career Napoleon could draw upon a deeper pool of French manpower—supplemented by foreign contingents—than his opponents. Maintaining unity of command in his own hands gave Napoleon an enormous advantage in his most successful phase.

Napoleon's aggressive temperament and his determination to force his opponents into a decisive battle explain why he preferred to take the offensive both strategically and tactically. A leading student of Napoleonic warfare, Gunther E. Rothenberg, states that in all of his battles Napoleon only stood on the defensive on three occasions (at Leipzig in 1813 and at La Rothière and Arcis in 1814) and then only after his initial attack had failed.[11]

Although Napoleon had made an impressive début as an independent commander in Italy in 1796–7, it was not until 1805 that his genius as a strategist was made fully apparent throughout Europe. On 23 August 1805 the *Grande Armée* left the Channel coast, crossed the Rhine, and advanced along the Danube to threaten the Austrian army's line of communications with Vienna. The advance body of the Austrian army at Ulm, under the command of General Mack, was enveloped and on 19 October 33,000 men surrendered without fighting a major battle. This was a rare example of a decisive victory achieved by manœuvre and almost without bloodshed. Napoleon then advanced beyond Vienna and destroyed the united Austrian and Russian forces at Austerlitz. By the terms of the Peace of Pressburg Austria was removed from the Third Coalition, ceded Venetia to France, and left France in command of Central Europe. In a matter of months a seemingly powerful alliance had been swept away by a unique combination of swift and ruthless operations and forceful diplomacy.[12] Austerlitz was one of the Emperor's two most complete victories, whether measured by relative casualties, psychological impact on the defeated armies, or decisive political outcome. This 'thunderstroke victory that destroyed the enemy army in a simple clash of arms, became almost every general's hoped-for means to the goal'.[13]

In the following year Napoleon achieved his most spectacular and decisive victory against Prussia in the battles of Jena-Auerstädt. When Prussia mobilized and advanced southward through Saxony, Napoleon took advantage of the *Grande Armée's* presence in southern Germany to assemble his forces between Bamberg and Würzburg before moving northward towards

Berlin, thereby threatening to intervene between the enemy and his capital. Napoleon left his communications with the Rhine unprotected, relying on superior numbers and mobility to decide the issue. In a classic example of the Napoleonic strategic approach, the *Grande Armée* of some 180,000 men was deployed in three columns of two corps each on a front 30 to 40 miles wide. By 12 October this huge 'battalion square' had moved around the left flank of the Prussians who were withdrawing northward between Weimar and Jena. On 13 October Napoleon wheeled the bulk of his forces westward against what he believed to be the main Prussian force at Jena, while sending Davout some 15 miles further north to strike at the enemy's rear. On the next day the two battles were fought with fronts reversed (the French facing west and the Prussians trying to break out eastward towards Berlin) but, contrary to French intelligence, Napoleon faced only a small part of the Prussian army, whereas Davout was attacked by the much larger main enemy force. As on some earlier occasions in Italy, Napoleon was rescued from a dangerous position by the excellent performance of his corps commander. Davout not only held up the attacking Prussians, but eventually forced them to retreat westward, where they mingled with the survivors of Napoleon's overwhelming victory at Jena. In perhaps the most vigorous and ruthless pursuit of the era the Prussian army was destroyed as a fighting force; the Prussian military system was utterly discredited and harsh peace terms were eventually imposed.[14]

The ramifications of this most crushing victory are instructive. Despite the annihilation of its main armies, Prussia kept up a desultory resistance, mainly in East Prussia and in isolated fortresses, for another eight months, thus hinting at future problems in translating victorious battles into acceptable peace terms in conditions of 'total war', where humiliation provokes patriotic resistance. Even more significantly, the catastrophic failure of the Frederician system led to a radical national reaction entailing wide-ranging military reforms, including the introduction of conscription, selection of officers on merit, and the creation of the first modern general-staff system. These measures, and other far-reaching reforms, transformed Prussia into a formidable military power by 1813.

There is no need here to consider Napoleon's later battles in similar detail. A partial exception, however, must be made for those at Aspern-Essling (22 May) and Wagram (5 July) 1809. The significance of the former was that, despite approximately equal casualties (23,400 Austrian to 21,000

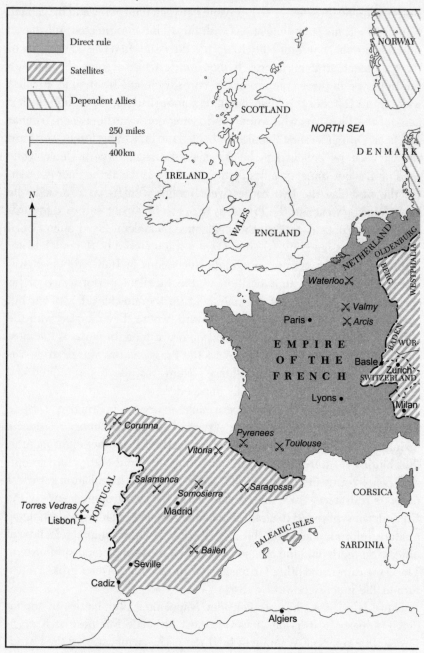

Map 2. French domination of Europe, 1792–1815

French), this was a clear victory for the Austrians under the Archduke Charles and the first definite personal defeat for Napoleon after his earlier indecisive 'draw' with the Russians at Eylau. Austrian military reforms had produced an army capable of beating the French, even if their able but circumspect commander failed to make the most of his brief opportunity of harrying the enemy in their disorganized crossing of the Danube. Aspern-Essling, following French defeats in Spain (notably at Baylen in 1808), marked a significant turning-point in Napoleon's fortunes.[15]

The improvement in the fighting capacity of the Austrian army was again manifest at Wagram on 5–6 July where its combat performance threatened the French with defeat until Napoleon mounted a brilliant counter-attack in which Macdonald's V Corps, strongly supported by artillery, split the Austrian army in two, and forced it to retreat. Each side lost approximately 40,000 men. The French army was badly shaken and its quality was beginning to suffer from such heavy losses.

This is not meant to suggest that the French army's decline was already clear to contemporaries in 1809; witness Austria's readiness to make peace after Wagram and Prussia's reluctance to renew the struggle until 1813. Nor could contemporaries detect any marked decline in Napoleon's generalship. On the contrary he conducted some of his most brilliant operations in 1813–14 when the numerical odds were stacked heavily against him. Indeed, his reputation and charisma remained so powerful that in the autumn of 1813 the allied war plan expressly advised retreat for any force against which he advanced in person.[16] What, however, may be suggested in the perspective afforded by hindsight is that, despite Napoleon's unrivalled abilities as a strategist and field commander, the period in which the French army's superiority was so marked as to produce truly decisive victories was quite brief.

Napoleon's reliance on short, relentless campaigns, culminating in a decisive battle and rounded off with an advantageous peace treaty, began to appear an increasingly hazardous gamble after his defeat of Russia followed by the Treaty of Tilsit in 1807. Since Napoleon never developed an adequate supply system and his armies grew ever larger, the latter were obliged to keep on the move and were heavily dependent first on capturing the enemy's magazines and then living off the produce and revenues of the defeated country.

In effect Napoleon wittingly placed himself on a treadmill of perpetual war in order to maintain his armies and imperial commitments. From 1805 onwards he systematically collected contributions and indemnities from occupied territories. After the defeat of Austria in 1805, for example, he raised 75 million francs from her territories, of which 48 million were spent in France. The campaigns in Prussia and Poland between 1806 and 1808 yielded about 482 million francs, of which 281 million returned to France; and the campaign in Austria in 1809 raised a further 164 million, of which about half returned to France. Quite apart from maintaining the bulk of his armies at his opponents' expense, Napoleon's conquests provided the French treasury with 10 to 15 per cent of its annual revenue from 1805 onwards.[17]

Clearly Napoleon was dependent on the disunity and self-centredness of his main continental opponents Austria, Prussia, and Russia, which would cause them to accept the verdict of battle and promptly sign peace treaties in the hope of eventually recovering their full independence. In the short term he was justified by such actions as Prussia's reckless, unsupported offensive in 1806 and her failure to support Austria in 1809; but Britain's continuing defiance and success in fostering new coalitions eventually led to his catastrophic defeat in 1812.[18] Long before that, however, the downward spiral in French fortunes began with large-scale military intervention in Spain after the popular revolt in 1808. In the first year of operations the Iberian campaign cost France 300 million francs and thereafter the annual costs steadily increased. The Russian campaign of 1812 had catastrophic consequences since it was far more expensive to mount than any other venture, brought no returns whatever, and left Napoleon deeply indebted to his contractors and provisioners.

Although the French army remained a formidable instrument of war until the end, it came to resemble a bludgeon rather than a rapier. There was a marked decline in the quality of French conscripts after 1806 and less attention was given to their training. Napoleon came to rely increasingly on foreign contingents whose motivation was dubious, particularly after the demoralizing experience of the 1812 campaign in Russia when only one-third of the 449,000 troops who crossed the Niemen on 24 June were French.[19] The declining quality of French troops was partly offset by increasing the number of guns and giving these a more prominent role. But still the predominantly attritional nature of the later battles brought heavy casualties. Crude French tactics at Borodino, for example, resulted in nearly

30,000 French casualties without destroying Kutusov's army. The three-day battle of Leipzig and the other battles of 1813, particularly in Spain, amounted to a 'decisive' defeat for France in attritional terms. At Leipzig the French suffered 45,000 battlefield casualties and lost a further 15,000 prisoners. The total French losses for the year were close to half a million— more than the nation could now afford. Meanwhile, Wellington was expelling the French armies from Portugal and Spain, thereby inflicting heavy losses in territory, manpower, and resources. Moreover, Germany was finally lost, so that henceforth Napoleon was fighting essentially in defence of the French frontiers.[20]

In these later campaigns Napoleon also paid a high price for his insistence on one-man rule, as regards both command and staff duties. He had never allowed his staff to develop responsibility for generating plans or exercising any real authority in the transmission of orders or the control of formations beyond his personal reach. As the French armies increased in size and their commanders acquired responsibilities in widely separated theatres of war (from Spain to Eastern Europe), Napoleon's strategic control collapsed.[21] Napoleon's failure to encourage his marshals to exercise individual responsibility brought serious consequences, particularly as some of his senior commanders were removed due to death or wounds, loss of energy, or political defection. In 1812 and all the subsequent campaigns there were instances where senior commanders failed to understand or carry out Napoleon's orders in critical situations. The failure to permit a genuinely autonomous theatre commander in Spain was only the most blatant example of a basic flaw in the Napoleonic system. The overextension of French military resources and command capacity at the extremities of Europe—in Spain and Russia—also revealed another, and fundamental flaw, in Napoleonic grand strategy; namely its over-reliance on achieving a quick decisive victory in battle. In Spain the guerrilla war, in inhospitable terrain and stiffened by Wellington's army, gave the overextended French forces no chance to deal a knock-out blow; while in Russia Napoleon's pursuit of the undefeated enemy armies to Moscow, and his sojourn there in the vain hope that the Tsar would negotiate, proved a reckless and, ultimately, fatal gamble.[22]

Finally, in adducing the military causes for France's decline and eventual defeat her opponents' improvements must also be taken into account.[23] Gradually, as France's enemies came to understand the ingredients of the former's successes, they strove to counter or emulate them so far as their military and political traditions permitted. Thus Austria and Prussia introduced forms of compulsory service and improved their tactics and military training. Prussia developed a general staff system which was far superior to any other at the time. In Britain, by contrast, it proved possible to implement significant military reforms within a highly conservative political and social context.

Taking a wider view of the changing balance of power between 1792 and 1815, Spenser Wilkinson surely captured an essential truth in his conclusion that:

Napoleon's fall repeats the conditions of his rise but with the roles exchanged. He had been created in the effort of a nation in arms against a group of absolute monarchies; he was undone by a group of nations in arms to upset his own absolute monarchy. Here was the political surprise that turned against him, for he failed to perceive that while he had been conquering Europe he had also been forcing upon its peoples the very conception of nationhood which had transformed France and given him his opportunity.[24]

In the period of his greatest triumphs Napoleon's combination of the roles of head of state and commander-in-chief had been one of his greatest assets, but in the longer term it was the fundamental cause of his failure. Unlike Frederick the Great, who had also enjoyed absolute authority as general and statesman, Napoleon proved unable to limit his ambitions. He could never 'bring himself to sacrifice certain ambitions for the better success of others. No mind was ever less capable of understanding the necessity of compromise.'[25]

His intense psychological need for conquest and absolute domination was not only antithetical to diplomacy and statesmanship: it also in practice limited him to the relentless pursuit of decisive victory in all-out war which Clausewitz found theoretically so admirable. As Paret comments, 'It is rare that a state's foreign policy stands in need only of major wars, yet Napoleon excluded limited wars for circumscribed goals from his political and military system'.[26] Short of the complete domination of Europe he therefore seemed doomed to provoke his opponents into making ever greater exertions—and to patch over their differences—until eventually a coalition of great powers would stick together long enough to bring about his absolute defeat.

Napoleon's megalomania also eventually exhausted the territorial resources of the European states system. In the 1790s—as the partitions of

Poland most vividly demonstrated—and even in the early 1800s there were opportunities for other states as well as France to expand or extend their indirect influence. There were spoils to be distributed in Italy, the Iberian peninsula, the Low Countries, Eastern Europe, and above all in the myriad of petty states and principalities of Central Europe. In time, however, Napoleon began to monopolize these territorial rewards, carving out kingdoms and dukedoms for his relatives and marshals, even to the point of altering the sovereign status of his satellites to provide for them. As late as 1807, when he signed the Treaty of Tilsit with the Tsar, Napoleon could still offer territorial gains to at least one other continental power, but by 1812 he had alienated virtually every state in Europe, including his former allies in Germany who now stood to gain from his downfall. Even in these dire circumstances Napoleon had reasonable chances of negotiating a compromise peace in 1813 and 1814 but preferred to fight militarily heroic but politically futile campaigns.[27]

Paul W. Schroeder goes to the heart of the matter in describing Napoleon's foreign policy as 'a criminal enterprise',[28] pointing out that the almost universal response in Europe to the experience of Napoleonic conquest was initially appeasement, submission, and efforts at accommodation—Prussia, for example, followed this unheroic course for a whole decade before 1806 and again after defeat in that year. Spain and Russia opted for uncompromising resistance in 1808 and 1812 respectively, only after enduring years of exploitation and humiliation as Napoleon's allies. The common conclusion from bitter experience of all France's opponents, including Britain, was that Napoleon could not be appeased. One state after another reluctantly resorted to war from the conviction that it was impossible to do business with the Emperor of France: that peace on his terms was more dangerous and humiliating than war. This applies for example to that most unbellicose sovereign, Emperor Francis of Austria, who once again risked war after three defeats *and without allies* in 1809. Francis, and other rulers such as Tsar Alexander, were forced to conclude that Napoleon could not be regarded as a normal statesman: he treated agreements like that made at Tilsit as scraps of paper and rode roughshod over his allies' interests, as when he demanded that the Tsar close his ports to British goods while he was trading openly with Britain in France's interest.[29] On this interpretation of his foreign policy, Napoleon could not draw a line under his conquests with a view to negotiating a lasting settlement because he had

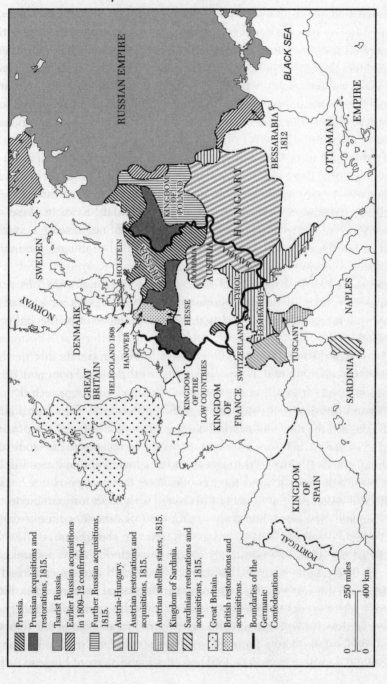

Map 3. Territorial settlement after the Napoleonic Wars

Prussia.

Prussian acquisitions and restorations, 1815.

Tsarist Russia.

Earlier Russian acquisitions in 1809–12 confirmed.

Further Russian acquisitions, 1815.

Austria-Hungary.

Austrian restorations and acquisitions, 1815.

Austrian satellite states, 1815.

Kingdom of Sardinia.

Sardinian restorations and acquisitions, 1815.

Great Britain.

British restorations and acquisitions.

Boundaries of the Germanic Confederation.

RUSSIAN EMPIRE

BLACK SEA

OTTOMAN EMPIRE

BESSARABIA 1812

KINGDOM OF POLAND

HUNGARY

PRUSSIA

BOHEMIA

AUSTRIA

BAVARIA

TIROL

SWEDEN

NORWAY

HOLSTEIN

DENMARK

HANOVER

HELIGOLAND 1808

GREAT BRITAIN

HESSE

LOMBARDY

SWITZERLAND

KINGDOM OF THE LOW COUNTRIES

KINGDOM OF FRANCE

TUSCANY

NAPLES

SARDINIA

KINGDOM OF SPAIN

PORTUGAL

250 miles

400 km

0

0

no final goal, no clear notion of a European system or international order. After fifteen years of Napoleonic conquest, domination, and exploitation, European statesmen and princes reluctantly concluded that an alternative to eighteenth-century politics had to be found because playing the old game with him was intolerable.[30] Lawless to the end himself, Napoleon's unintentional achievement was to force Europe after 1815 to adopt a new system of international politics based on co-operation of the *status quo* powers and a greater respect for law.

In the short term the concert of the victorious powers of 1814 and 1815 was primarily designed to prevent the resurgence of a revolutionary, aggressive France while preparing her for a speedy readmission to the club of great powers. In the longer perspective it became clear that national exhaustion resulting from Napoleon's prolonged and intensive warfare had ended French prospects of military, naval, and economic dominance. Despite a period of great hardship and war weariness after 1810 (exacerbated by war with the United States), Britain emerged from the Napoleonic wars with her overseas empire extended, with a strong navy and mercantile fleet, and with an increasing lead in the industrial revolution.[31]

In view of the failure of Napoleon's most brilliant and militarily 'decisive' victories to do more than temporarily eclipse his principal opponents,[32] the attritional and extremely costly nature of his later campaigns including the disastrous invasion of Russia), the frightful implications of the guerrilla war in Spain, and the final unequivocal defeats on or close to French soil in 1814 and 1815, it is at first glance surprising that the heroic image of Napoleon as military genius flourished throughout the nineteenth century and well into the twentieth. But such was the case. As Peter Paret suggests in a brilliant reflective passage,[33] the impact of his reign, with its almost continuous run of dramatic campaigns, had been so profound that the sequence of defeats in the final phase did little to reduce his towering stature; indeed, the very fact that the allies had eventually beaten him after so many failures may, paradoxically, have enhanced his status. Moreover, Napoleon embodied the wild, romantic spirit of the early nineteenth century, which contrasted so starkly with the cool and moderate rationalism of the pre-revolutionary era. Short of dying in battle like Nelson or Moore, Napoleon's heroic failure at Waterloo, followed by a terminal exile on the island of St Helena, was just about the next best apotheosis for laying the foundations of a legend. Would the legend have been quite so powerful had he lived in comfortable retirement for a further thirty years to equal the life span of his exact contemporary Wellington?

Furthermore, though eventually pushed beyond its effective limits by Napoleon's boundless drive and ambitions, the form of 'total' warfare which he exemplified in the 1800s did indeed exploit to the full the technological, logistical, and manpower conditions of the time. Studious professional officers tended to overlook the failure of Napoleon the statesman in their preoccupation with the remarkable achievements of Napoleon the general. The essential components of his generalship seemed to consist of mobilizing superior numerical strength, carefully planning and implementing deep strategic penetration to disrupt and disorganize the enemy's deployment, and rapid concentration of force at the decisive point in order to achieve a total victory which would destroy the enemy's army and so end the war. Thus Napoleon demonstrated repeatedly that, as a military commander, he could win 'decisive victories' on the battlefield, but as a statesman he failed abysmally to translate these successes into enduring political victories. It is with the theoretical development of the notion of Napoleon, in the role of military commander, as the outstanding exponent of the decisive battle, that we shall be mainly concerned in the next chapter.

The Napoleonic Legacy:
The Influence of Jomini and Clausewitz

Of the numerous historians, military theorists, and polemicists who in the nineteenth century contributed to the legend of Napoleon as the 'God of War' the two most influential were surely the Swiss Antoine-Henri Jomini (1779–1869) and the Prussian Carl von Clausewitz (1780–1831). Although there were many contrasts in their careers as professional soldiers, their political and cultural outlooks, and their approaches to the study of war, Azar Gat argues persuasively that the recent tendency to put them at opposite poles as interpreters of Napoleonic warfare is somewhat exaggerated. Both in fact 'reflected the spirit and outline of Napoleonic strategy'.[1]

Although his early writings were influenced by the military theorists of the Enlightenment, notably Henry Lloyd, Jomini quickly appreciated the transformation in warfare effected by the French revolutionary armies in the 1790s, and as a young officer he served on the staffs of Ney and Napoleon himself in the Ulm, Jena, Eylau, Spanish, and Russian campaigns. As regards the destruction of the enemy's field armies in decisive battle, Jomini was still quite conservative in outlook in 1804–5, contending that battle should only be risked when conditions were very favourable. But, like Clausewitz his tone and emphasis became much bolder after Napoleon's decisive campaigns of 1805–7. Now, he argued, the purpose of moving upon the enemy's line of communications was to bring him to battle, after which he should be pursued relentlessly. He appreciated that seizing the initiative and employing rapid, flexible movements were vital features in Napoleonic strategy. Most important of all was the concentration of force; the employment of masses against the decisive points.

From the outset Jomini was concerned to discover and formulate permanently valid principles based mainly upon Napoleonic operations. In essence he deduced that seizing the initiative and making forced marches should be used to concentrate maximum force and achieve local superiority at the decisive point. While he avoided any attempt at the mathematical precision or the pedantry of some of his contemporaries, Jomini always believed that the real secret of battlefield success lay in the skilful use of the lines of operations.[2] He won fame while still a young man by combining an updated version of the operational system of the Frederician era with the more aggressive spirit and style of Napoleonic warfare. He stressed the desirability of overthrowing the enemy's army, either through rapid manœuvres to cut his communications, or by penetrating and overwhelming his divided forces in turn from a central position; that is, the alternative strategies of envelopment or the use of interior lines.

Thus Jomini was more overtly prescriptive in approach than Clausewitz: the former was primarily concerned with the establishment of principles derived from a study of operational movements; the latter with discovering the nature of war. Both laid more stress on the significance of the decisive battle than their predecessors, but Jomini's cool analysis and didactic attitude lacked the passionate, personal commitment which enlivened Clausewitz's extended and almost bloodthirsty discussions of *die Schlacht* (meaning 'the battle' but also 'slaughter'). As Michael Howard observes, the emotional quality of the writing suggests that 'he regarded a campaign that culminated in such an encounter as somehow morally superior to one that did not'.[3]

Clausewitz held consistently to the view that the direct annihilation of the enemy's forces must always be the overriding consideration. Such destruction can usually only be achieved by hard fighting (bloodless victories like Ulm are exceptional, he warned, and must never be counted on); those major engagements which involve all available forces lead to significant victories; and the greatest victories result where all the engagements coalesce into one great battle.[4]

Although Clausewitz was not 'prescriptive' in a dogmatic sense, he clearly believed that history provided lessons which the general would neglect at his peril. Between the earliest drafts of 1804 and the crisis he experienced in 1827, the 'lessons' could be summarized as follows: aim for great objectives, achieve the maximum concentration of force, act as aggressively as possible to destroy the enemy army in a decisive battle, and finish off the enemy state's capacity for resistance by a ruthless pursuit. He was contemptuous of commanders who evaded decision in battle or put too much emphasis on manœuvre—they were acting in contradiction of the spirit of war. Clausewitz focused his attention single-mindedly on battle (or 'the engagement'). The role of strategy was to secure superior numbers and a concentration of force at the decisive point. He placed little value on strategems or surprise and virtually excluded from his theory considerations of supply and maintenance—to which Jomini gave great emphasis. In his treatise *Principles of War for the Crown Prince* (1812) he made the astonishing remark that 'Napoleon never engaged in strategic envelopment'.[5] Consequently, as later critics were to show, he was an inadequate guide to Napoleon's strategic methods or principles, and in this respect inferior to Jomini.

Clausewitz profoundly admired Napoleon as a general but spent the best years of his professional life in fighting the French, even transferring to Russian service in 1812 at great personal cost in order to continue the patriotic struggle. Jomini, by contrast, spent his most professionally rewarding years in French service (being promoted a colonel on Ney's staff as early as 1805) before he too transferred to Russian service in 1813 for more personal reasons than the Prussian. In old age Jomini doubtless exaggerated his personal association with, and influence on, Napoleon but he did win the Emperor's praise for his early contributions to military theory, while on St Helena Napoleon would display a remarkable empathy with Jomini's historical and theoretical ideas.[6]

Jomini inverted a fundamental eighteenth-century assumption about the study of war by arguing that, whereas tactics were constantly changing and so difficult to regulate, strategy could be reduced to universal, enduring principles. By leaving aside much of war's social and political context he was able to bring great clarity to the analysis of operational planning, manœuvres, and results. Critics would contend that he turned war 'into a huge game of chess', but his approach proved extremely popular over a long period—particularly in military academies. Indeed, John Shy proposes the qualified compliment that he, rather than Clausewitz, 'deserves the dubious title of founder of modern strategy'.[7]

Part of Jomini's appeal as a prolific historian, particularly in the aftermath of Waterloo, was that his schematic interpretation of recent events provided reassurance and certainty after a generation of turmoil, confusion, and destruction. His main proposition was that the revolutionary generals, and, even more, Napoleon, had owed their successes to the practice of correct strategic principles and had eventually failed through neglecting them. The implication was that the conduct of war was a science which could be assimilated through study.

Jomini's theory enunciated as early as 1803 was that strategy is the key to warfare; that strategy is controlled by unchanging scientific principles; and that these prescribe offensive action to bring superior mass forces into action against a numerically weaker enemy at some decisive point in order to secure victory.[8] He was deeply impressed by the new style of warfare in the 1790s which was reckless of manpower and the constraints of supply, which took risks and put maximum energy into the quest for victory in battle. While allowing Frederick the Great some credit for his outstanding victories, Jomini argued these were due to tactical skills because military thought had not yet formulated the strategic principles perfected by Napoleon. He even presumed to remark that Frederick 'entirely misunderstood' the principles of operations.[9]

At the height of Napoleon's military triumphs Jomini's summary of strategic principles was so broad as to seem to guarantee the Emperor victory, whether he operated against his enemy's flanks and rear to cut him off from his base, or pierced his opponent's centre so as to defeat his separated forces in detail. According to Jomini, Napoleon's genius lay in his unrivalled skill in exploiting interior lines so as to make the most of the central position in battle.

Napoleon's successive defeats in the campaigns of 1813, 1814, and 1815 caused Jomini much embarrassment. In 1813 the allies had shown the advantage of operating on exterior lines to envelop the French, thereby severing the latter's communications with their base of operations. Napoleon's skilful manœuvres failed either to defeat the allies separately or prevent them from concentrating on the battlefield at Leipzig. In the 1814 operations, Napoleon was at his brilliant best but his rapid marches and optimum use of the central position still did not permit him to overcome superior numbers. Finally, in 1815, Napoleon's swift advance designed to break through the allied front and defeat their separated forces in turn nearly succeeded but was thwarted by French command errors and the allies' determination not to be defeated piecemeal. In particular Blücher had sealed the allied victory by advancing on interior stategic lines. Here lay the fundamental problem for all system makers: if strategy was a science whose principles could be learnt what was to prevent all the belligerents learning them? In that case stalemate or attrition must result.[10]

Jomini was put on to the defensive, and was obliged to modify, though not to abandon, his faith in the central position and interior lines. The limited wars of the 1840s and 1850s could be accommodated to his theory but later 'exceptions', such as the Austro-Prussian War of 1866, would prove to be so significant as to cast doubt on the very basis of the principle. Other flaws in his approach and methodology became clear to careful readers of his works: he was, for example, reluctant to examine historical cases which seemed to contravert his principles; and he was ambiguous about the 'decisive point' and how it could be located. In the supplement to his Summary, for example, he wrote that: 'It is almost always easy to determine the decisive point of a field of battle, but not so with the decisive moment; and it is precisely here that genius and experience are every thing, and mere theory of little value.'[11]

Although Jomini's overriding concern was to reduce the complexities and chaos of warfare to a minimum with a view to discovering, and promulgating, permanently valid principles, he was by no means unaffected by the novel intensity, rapidity, and drama of Napoleon's campaigns which he had experienced. It is, moreover, to his credit that he perceived the darker, tragic side of Napoleon's audacious and ruthless marches. The art of war may have been enlarged by this system, but humanity had lost by it. The rapid incursions of mass armies feeding upon the regions they overran were not materially different 'from the barbarian hordes between the fourth and thirteenth centuries'.

In the conclusions to his Summary, Jomini acknowledged that the conduct of war could be significantly affected by the passions and morale of the antagonists, while chance or superior leadership could prevail in battle over strict adherence to the rules. Nevertheless, he insisted, there *were* permanent rules governing both tactics and strategy:

It is true that theories cannot teach men with mathematical precision what they should do in every possible case; but it is also certain that they will always point out the errors which should be avoided . . . for these rules become, in the hands of skilful generals commanding brave troops, means of almost certain success.

He could only pity prejudiced officers who read his work and still doubted that there are principles and rules; they were like Frederick the Great's mule which was none the wiser after twenty campaigns. Towards the end of his long life Jomini recognized that 'the means of destruction are approaching perfection with frightful rapidity'. The revolution in fire-power would profoundly affect organization, armament, and tactics. Yet 'Strategy alone will remain unaltered, with its principles the same as under the Scipios and Caesars, Frederick and Napoleon, since they are independent of the nature of the arms and the organization of troops'.[12]

Gradually the technological and political transformation of the nature of war would make Jomini's orderly approach, with its stress on lines of operations and decisive points, appear too restricted and formal. But as generations of military historians during the nineteenth and first half of the twentieth century discovered, it was easier—and more profitable—to write Jominian rather than Clausewitzian volumes. It was also much easier to *read* Jomini in French than Clausewitz in German—or in defective English translations. Jomini had indeed constructed the clearest and most practical framework for the study and teaching of Napoleonic and any other warfare. The majority of cadets and officers in North America as well as Europe would receive their knowledge of military history, and generalship in particular, through various editions of Jomini's *Summary of the Art of War*. In the operational sphere the Napoleonic model also dominated Russian military thought until 1914. As communications, weapons technology, and the political and social components of war began to change drastically, Jomini's highly influential legacy contributed to the dangerous 'Napoleonic' delusion that superior strategy and tactics in the decisive battle would assure victory, irrespective of the political and social context in which the battle was fought. It was, after all, the general's overriding responsibility to secure victory on the battlefield and the statesman's to control policy, but it would need their combined judgements—and wisdom—to decide whether such a goal was realistic in particular circumstances.[13]

It is easy to understand why Jomini's fame continued to grow through the middle decades of the nineteenth century while that of his rival, Clausewitz, never a popular author during his lifetime, suffered an eclipse after his sudden death from cholera in 1831, despite his wife's posthumous publication of his works, including *On War*, in the following decade. It was not just that Clausewitz did not survive to complete his most famous work or respond to criticism. The high tide of German Romanticism also began to ebb and with it some of the preoccupations of the Napoleonic era, such as the heroic struggle against the French and its domestic corollary, the enhancement of Prussian security and prestige by encouraging more popular participation in politics and the armed forces. In the short term Clausewitz's speculation that the final overthrow of Napoleon might be followed by a period of limited, inconclusive wars, reminiscent of the eighteenth century, proved accurate. Jomini was much better attuned, in his detached, confident, and formalistic approach to war, to the unheroic decades of the 1840s and 1850s. Residing mostly in Paris, though continuing to advise the Tsar—and indeed helping to establish a new Russian military academy—he continued to publish voluminously, including a multi-volume history of the French Revolution and a four-volume biography of Napoleon. An expanded two-volume version of an earlier *Synoptic Analysis of the Art of War* was published in 1837–8 as the *Summary of the Art of War*. This was immediately popular and became his most famous and most often translated study.[14]

Although John Shy remarks tartly that 'in the last fifty-six years of his life (i.e. until his death in 1869) there is surprisingly little intellectual development', this did not prevent the spread of his influence. One problem which did worry him was how to incorporate into his 'art of war' a strategy for dealing with guerrilla, partisan, or people's war, the ferocity, horrors, and disorganized nature of which he had witnessed in both Spain and Russia. As a soldier Jomini preferred 'loyal and chivalrous warfare' to organized assassination, and the 'good old times' when the French and English Guards courteously invited each other to fire first to the 'frightful epoch when priests, women and children throughout Spain plotted the murder of isolated soldiers'. Thus, his view was that such wars were best avoided: they were too costly, too destructive, and too uncontrollable to be assimilated into his orderly elaboration of strategic principles.[15]

Jomini's direct influence through his publications and his indirect influence through eminent followers such as Dennis Hart Mahan in the United States and Sir Edward Hamley in Britain, make him the most fashionable military pundit in those two countries in the mid-nineteenth century, as well as in his adopted homeland, France. Jomini's influence so permeated the teaching of strategy of West Point in the 1840s and 1850s that a modern historian remarked with little exaggeration that the generals on both sides in the American Civil war went on campaign with a sword in one hand and Jomini's *Summary of the Art of War* in the other.[16] The precise nature of that influence has been much debated because by the 1860s tactics were being dramatically affected by enhanced fire-power and strategy by the development of railways and the telegraph. Even in Germany during his own long lifetime Jomini's influence at least equalled Clausewitz's. As Shy points out, it is an egregious error to imagine the majority of Prussian officers before 1870 poring over *On War* and treating it as their bible. Most of the German versions were poorly edited and difficult to read, particularly by impatient officers seeking authoritative operational principles. Instead, many turned to ardent Jominians such as Wilhelm von Willisen, whose *Theory of Great War* was published in 1840.[17]

After 1870, as will be discussed in the next chapter, Clausewitz's name did indeed become universally known as the symbol (and partial source) of German military prowess, but as Shy perceptively comments, Jomini had already won their personal duel by conditioning generations of officers to approach the study of war on his terms. Consequently Clausewitz's *On War* was considered first and foremost as an operational manual. Not surprisingly, Clausewitz seemed to add little to Jomini's familiar teachings as the authoritative interpreter of Napoleonic warfare, while his concern with such matters as the unpredictable element in war, the limited use of theory, and the notion that war was essentially a continuation of politics not an autonomous activity, were not particularly welcome to his professional readers. Above all they found unacceptable his clear statement that the defensive was theoretically stronger than the offensive. In Shy's apt phrase, Jomini's books had the effect of 'desensitizing their audience to the vital parts of Clausewitz's message'.[18]

If, however, Jomini had the advantage over Clausewitz in influencing the education of generations of officers in operational strategic concepts as distilled in the principles of war, Clausewitz made an important 'comeback', especially after 1870, on a wider range of issues concerning the nature of war, the morale of armies, and the qualities to be looked for in military leaders. Most importantly, Schlieffen and other senior German officers praised *On War* for 'its emphatic accentuation of the annihilation idea'.

In contrast to Jomini's clear but limited scope, which admirably catered for the needs of non-intellectual officers, Clausewitz offered a more elevated vision of the significance of war in international politics which would appeal to a more varied readership in the euphoric, nationalistic atmosphere after German unification.

Few students of military history and strategy in the late twentieth century dispute Clausewitz's status as the outstanding theorist of war.[19] In the past twenty years or so his career, works, and influence have been the subject of so much excellent scholarship that there is no need to attempt a summary or extended commentary here. What has emerged from the work of scholars such as Peter Paret, Sir Michael Howard, W. B. Gallie,[20] and, above all, Azar Gat is that *On War* is a difficult work to understand, and not only because its author died without revising more than a fraction of it, thereby creating confusion in the very first chapter with its emphasis on absolute war. Beyond that there are fundamental inconsistencies and even contradictions in the methodology which can now be clarified and discussed, though not of course resolved. Recent scholarship, notably by Peter Paret and Azar Gat, has thrown fresh light not only on the evolution of the manuscripts that have come down to us as *On War*, but also on the philosophic and cultural contexts in which Clausewitz prepared his ambitious work. Perhaps the outstanding result of these labours—apart from the important one of making the text of *On War* more accessible to the serious but non-specialist student—is to establish even more clearly than before the author's achievements—and limitations. He was, after all, a professional soldier, albeit with a very unusual mind and personality, whose adult life was encompassed by the period 1801–31.

One of the most important points to emerge from Azar Gat's revisionist reappraisal of the development of Clausewitz's thinking about war is that from the earliest note of 1804 through to 1827 he consistently and even vehemently stressed the supreme importance of 'absolute war' as an ideal; that is the need for a massive concentration of forces and an aggressive strategy aiming at the total overthrow of the enemy.[21] This was the lesson he had drawn from the allies' half-hearted conduct of the war against revolutionary France in the 1790s which had led to Prussia's defeat in 1795.

His convictions were enormously strengthened by his personal experience in the national defeat and subsequent humiliation at Napoleon's hands in 1806. Prussia had clung to the essentials of Frederician military organization, tactics, and command system until these catastrophic defeats (at Jena-Auerstedt) suddenly discredited the entire political and social foundations of eighteenth-century diplomacy and warfare. To counter Napoleon, Prussia needed to implement far-reaching military and political reforms designed to secure a much greater mobilization of manpower and resources, an injection of patriotic fervour, and an aggressive strategy to bring about rapid decision in battle. As a junior officer, Clausewitz played only a minor role in the post-1806 reform movement, and he was only one of many studious, patriotic officers who were shocked by the events of that year. But no one gave these emotions and ideals more impassioned expression than Clausewitz. As Michael Howard noted, 'Clausewitz's reasoning is as flawless as his passion is understandable . . . no one who had experienced Napoleonic warfare could have quarrelled with his statement "the character of battle is slaughter, and its price is blood"'.[22] This was the horrible reality of warfare which Bülow, Jomini, and other strategic analysts of the day tended to underplay if not ignore.

Since Clausewitz was profoundly impressed at a crucial stage in his intellectual development by the revolutionary novelty of 'total' Napoleonic warfare, he was evidently well aware of the character of the indecisive, restrained, and—in later historians' terms—'limited' warfare which it superseded. Indeed, in several emotionally charged passages he criticizes the pusillanimity of many eighteenth-century commanders for their less than total commitment to achieving great objectives through decisive battle. As a corollary he must have perceived that indecisive warfare was significantly influenced by political considerations, but until very late in his life he did not pursue this point to its natural conclusion, namely, that modest policy objectives would logically result in limited wars. His immediate practical response—under the shadow of Jena—would be that it takes two sides to keep war limited and in present circumstances anything less than a total effort would risk national disaster and even, as the case of Poland illustrated, extinction.

Azar Gat and Jan Willem Honig have demonstrated that, until his post-1827 revision, Clausewitz did not employ the term 'limited war' and, furthermore, did not recognize it as a legitimate option.[23] In 1804 he defined the purpose of war as either to destroy the enemy's state or to dictate the terms of peace. Both options were mentioned in 1827 under the single aim of overthrowing the enemy. Similarly in 1812 he listed the complementary aims in the conduct of war as to conquer and destroy the armed forces of the enemy, to take possession of his military resources, and to win public opinion. This repeated stress on the 'total' Napoleonic style of warfare aiming at the destruction of the enemy forces by means of the great decisive battle found expression in the early and unrevised draft of *On War* which comprises books ii–vii inclusive, the greater part of the work.

Clausewitz scholars have long debated the order of composition and significance of the two 'final notes' because they have a bearing on the nature of the far-reaching revision of *On War* which the author was contemplating in 1827. Azar Gat has provided the most comprehensive analysis of these documents in relation to the entire history of the posthumously published work.[24] Clausewitz experienced a crisis concerning his whole philosophy of war and methodology when he belatedly confronted the fact that the great weight of historical evidence contradicted the prescriptive priority he had given to all-out, 'total' war as outlined above. It seems likely that this intellectual breakthrough resulted from the author's gradual change of view about the new priorities in Prussian foreign and military policies after the end of the era of heroic struggle against Napoleon. Prussia's needs in the 1820s had a good deal in common with those of the 1780s.

It was in discussing 'the defence of a theatre of operations' that respect for historical evidence forced him to admit that: 'History records numerous cases that do not lack for an aggressor or a positive ambition on one side at least, but where this ambition is not pronounced enough to be relentlessly pursued until it leads to the *inevitable* decision.' In other words, neither side is willing to seek a decisive battle. This insight led him to reflect that:

The history of war, in every age and country, shows not only that most campaigns are of this type, but that the majority is so overwhelming as to make all other campaigns seem more like exceptions to the rule. Even if this ratio changes in future, it is certain that there will always be a substantial number of campaigns of this kind, and that aspect must have its due in any doctrine of defending a theater of operations.[25]

In short, the Napoleonic style of 'total war', which, he had argued, expressed the true nature of war, was shown to be highly exceptional. Historical evidence pointed to the incontrovertible conclusion that the vast majority of conflicts were not struggles for life and death; if anything they approximated to a mere state of armed observation. Clausewitz admitted to himself that the very foundations of his conceptual framework of war were threatened, in that his theory based on logic was at odds with historical reality.[26]

To a modern mind, and particularly a pragmatic Anglo-Saxon one deeply suspicious of philosophical idealism and concepts such as absolute war, the solution to the dilemma would seem to be simple. Since, as Clausewitz acutely recognized, the nature of war was largely determined by its political and social milieu, there should be no difficulty in accepting that Napoleonic warfare was a reflection or embodiment of the special conditions prevailing in France in the 1790s and 1800s. If wars were essentially political instruments they would legitimately cover the whole spectrum, from minor frontier clashes to all-out struggles for annihilation or survival.

Clausewitz, however, was very reluctant to abandon the concept of the 'total' war of destruction as the ideal or norm against which all lesser forms should be measured. He therefore tried to bridge the gulf between theory and reality by recognizing two types of war, while claiming that the 'total' war of destruction represented the true nature of war and must take priority. His reasoning was that against a half-hearted war effort an all-out one will always prevail. Limited war, he argued, was not a genuine form of war but one which was modified by various inhibiting factors, internal and external. These factors were alien to the true nature of war and served to limit its intensity.[27] Among these 'external factors' the most important was the degree of control exercised by policy. This point is developed in the often-quoted chapter entitled 'War is an Instrument of Policy'. But, contrary to the late twentieth-century impulse to interpret this in a sense favourable to democracies at war, Clausewitz expressly states that the influence of politics is an *external* force, which works *against* the true essence of war. Policy, in effect, 'converts the overwhelmingly destructive element of war into a mere instrument'.[28]

For obvious reasons Clausewitz was not satisfied with this bridging operation because he appreciated that, if the nature of war was strongly influenced by political considerations, then the primacy given to absolute war lost much of its point. In writing book viii and revising the early chapters of book i Clausewitz moved a long way towards accepting the validity of history in all its variety as against enshrining absolute war as the ideal. Essentially, in these sections written or revised towards the end of his life, Clausewitz now allowed equal status to a variety of war aims and operational objectives while continuing to regard the clash of forces in battle as the most important and reliable means for attaining any political goal. Some political objectives do not require the opponent's outright defeat, but the *total use* of the *means available* must remain the best option: 'Kind-hearted people might of course think there was some ingenious way to disarm or defeat an enemy without too much bloodshed, and might imagine that this is the true goal of the art of war. Pleasant as it sounds, it is a fallacy that must be exposed.'[29]

Whether Clausewitz would have completely resolved the tensions and contradictions remaining in the text of *On War* had he lived longer will never be known. Given the author's ruthless self-criticism, concern for the truth, and respect for objective historical evidence, it is possible to assume, as does Michael Howard, a positive answer. Azar Gat and Gallie, however, remain sceptical because Clausewitz was most reluctant to abandon the concept of absolute war as an ideal and seems to have satisfied his own doubts by adopting a Hegelian device of retaining contradictory components but viewing them from a higher standpoint, on the curious assumption that 'all the contrasts and contradictions of reality were actually but differing aspects of a single unity'.[30]

Be that as it may, the common-sense conclusion is that there were stronger grounds than it is now customary to admit for later nineteenth century nationalists to find in *On War* the classic exposition of their belief in the prime importance of the great battle. They were closer to Clausewitz in their view of the nature and efficacy of war than most post-1945 commentators, who discovered in his work 'lessons' appropriate to their concerns with deterrence and limited war in the nuclear age.[31]

It should come as no surprise to modern students, in the light of Peter Paret's admirable study *Clausewitz and the State*, that Clausewitz, a lifelong professional Prussian officer from the age of 12, was profoundly influenced by the cultural and political ethos of his youth and early manhood. In other words, while it would be unjust to label him a 'militarist', he does occasionally express views which make the modern reader feel uncomfortable. For example:

Today practically no means other than war will educate a people in this spirit of boldness; and it has to be a war waged under daring leadership. Nothing else will counteract the softness and the desire for ease which debase the people in times of growing prosperity and increasing trade.

A people and nation can hope for a strong position in the world only if national character and familiarity with war fortify each other by continual interaction.[32]

But it is equally anachronistic to credit him with the tenets or outlook compatible with a late twentieth-century popular democracy. Clausewitz was a liberal only in that he welcomed the abolition of the legal foundations of the rigid pre-1807 class structure and favoured opening the business and military professions to talent. But he was not a democrat under any common-sense definition; rather a tough-minded paternalist, deeply influenced by Machiavelli, who took an exalted, idealist view of the state and its authority *vis-à-vis* the individual. Clausewitz, however, was more concerned than Machiavelli with the war-making role of the state as the embodiment of the community's general interests, *towards other states*. This, as Gallie points out, expresses his recognition of the all-important truth that it makes little sense to talk of 'the state' *per se*: 'no state would be a state if it did not exist as one of a plurality of other and (at least potentially) rival states.'[33]

Finally, in view of Clausewitz's open defiance of his king and government in 1812 on the grounds that he (and some 200 like-minded officers) had a higher patriotic conception of their duty than their rulers, and that their view of the national interest obliged them to transfer to a foreign power's (i.e. Russian) service, there is some irony to be savoured in the belief that he bequeathed a useful model for the practice of civil–military relations.[34]

This chapter has suggested that, although Jomini and Clausewitz were quite similar in their early reactions to and interpretation of Napoleonic warfare, the post-1815 trajectories in their professional careers, intellectual development, publications, and international reputation were markedly different. Jomini had the tremendous advantage of living until 1869, thus surviving most of his enemies, critics, and rivals, against many of whom he had the proverbial 'last word'. He claimed Napoleon's approval, enjoyed a successful second career in Russian service, published an enormous amount, and became the predominant influence in military (and naval) education in several countries, including the United States and Britain.

On these criteria Clausewitz's post-Waterloo career was a disappointing anticlimax. His service prospects never fully recovered from his defection to Russia. His attempt to leave the unexciting administrative directorship of the *Kriegsakademie* by becoming the Prussian ambassador in London was frustrated. He published comparatively little in his lifetime and was not well-known even in Germany. His return to active service in Poland in 1830 was saddened by the sudden death of his revered commander, Gneisenau, and he died shortly afterwards; both were victims of cholera. His wife's great effort in publishing his works posthumously had only a short-term impact and by the 1860s his reputation 'seemed to have fallen into respectful oblivion'.[35] Moltke's sensational victories in 1866 and 1870 would effect a remarkable transformation.

Moltke and the Wars of German Unification

Prussia's sensational victories in 1866 and 1870–1 are pivotal to any consideration of the pursuit of victory in modern history. In purely military terms they suggested that decisive, Napoleon-style campaigns were still possible in the railway age and thereby created a seductive but dangerous model. Even more impressive than Napoleon's campaigns, they were near-perfect examples of Clausewitz's celebrated axiom that war is an instrument of policy. As this chapter will argue, it was mainly due to Bismarck's state-craft that these wars were politically controlled and yielded lasting political gains to the victor. Austria never challenged her expulsion from Germany and the settlement with France lasted until 1914—even then Alsace-Lorraine was not the *casus belli* and it was Germany who invaded France. The greatest prize, German unification, seemed to have been achieved at a very reasonable price and would only be undone some seventy-five years later by reckless adventures undreamt of in the Bismarck era.

The 1860s truly constituted a watershed in military history. Contemporary critics and students of war quickly perceived the essential elements which promised to transform the conduct of military operations. The American Civil War and Prussia's wars displayed a marked increase in the number of combatants since the peak reached in the Napoleonic era, a dramatic increase in fire-power, and the impact of the industrial revolution on transport and communications as well as weapons. Greater popular involvement, encouraged by nationalist fervour and embodied in forms of compulsory service, threatened even more far-reaching changes, already foreshadowed by Clausewitz in his discussion of total war, but not fully demonstrated until the First World War.

The political structure and the system of international relations estab-lished at the Congress of Vienna in 1815 and subsequent conferences of the great powers had endured remarkably well. The map of Europe had been redrawn on the grounds of security and to achieve a balance of power among the monarchist victors over Napoleon, who soon readmitted France to their ranks. Thenceforth the Concert of the Powers aimed to preserve the *status quo* as regards state boundaries, while preventing or suppressing nationalist movements and revolution within their empires. The post-1815 system was notable for a new interventionist doctrine to preserve the balance of power, from which Britain soon excused herself. Although it failed to prevent endemic conflict in peripheral areas such as the Balkans and the Iberian peninsula, the collaboration of the great powers after 1815 was remarkably successful in creating a zone of peace in Central and Western Europe lasting precisely a century. There were only four wars among the peacemakers of Paris and Vienna, and of these only the Franco-Prussian War was sufficiently long and intensive to leave a residue of hostility. Even in that conflict it should be remembered that the worst slaughter was inflicted by Frenchmen on their fellow citizens in the Commune.[1]

Within this political system which prized stability and generally ignored or repressed popular aspirations for change, armies everywhere reverted to an institutional character and role reminiscent of the eighteenth century. Standards of military efficiency which depended on significant changes in society were deliberately neglected: 'Whatever value Napoleonic warfare might have for a power trying to overthrow the states system of Europe, it could have little for statesmen trying to preserve it.'[2] This cautious, calculating spirit, accompanied by a good deal of stagnation in the armed forces, characterized the low level of military preparedness and limited warfare of the half-century after Waterloo. The conflicts of the 1860s consequently came as a shock to a generation which had not only forgotten the intensity of Napoleonic warfare, but also had good reasons to believe that industrial and commercial developments were consolidating an era of peace and cooperation.

Nowhere was the shock greater than in Britain and the United States where geographical security reinforced the pacifistic optimism that liberalism and free trade would diminish the causes of international conflict. It is nevertheless surprising that a best-selling classic of popular military history, directly relevant to our theme, should have been published in Britain as early as 1851. This was Sir Edward Creasy's *Fifteen Decisive Battles of the World: From Marathon to Waterloo*. The author was an Eton and Cambridge-educated barrister, later the Chief Justice of Ceylon, a prolific amateur historian and, briefly, holder of a chair of history at London University. Creasy, though certainly Eurocentric in his selection of battles, was not naïve enough to suggest that decisive battles had ended with Waterloo—he hinted, for example, that Russian autocracy was a suitable case for the appropriate military treatment—but he did argue vigorously that all the battles described had contributed to the glories of European (and specifically) Victorian civilization. In other words, war had had a beneficent purpose: it had made the nineteenth century. It was, in John Keegan's striking phrase, 'the Whig interpretation of history writ in blood'.[3] Creasy concluded his book with a joyful vision of the 'banners of every civilised nation waving over the arena of our competition with each other, in the arts that minister to our race's support and happiness, and not to its suffering and destruction'. England was now setting a peaceful example 'in the great cause of the general promotion of the industry and welfare of mankind'.[4]

Creasy's magisterial survey of the positive results of great, decisive battles clearly caught the popular mood. By 1854 *Decisive Battles* was already in its fifth edition and by 1894 it had been republished no less than thirty-eight times. By then Creasy had had British and American imitators, the latter including Thomas Knox's *Decisive Battles since Waterloo* (1887); and the indecisive character of numerous world-wide conflicts since 1914–18 have done nothing to stem the flow. On the contrary, the concept of the 'decisive battle' or war continues to have irresistible appeal to authors, publishers, and readers alike. In the inter-war period, for example, Liddell Hart and J. F. C. Fuller both boosted this approach to military history, the latter most notably in his separate two-and three-volume series on *Decisive Battles of the Western World*.[5] Even German historiography has been influenced by the Creasy tradition as exemplified by two post-mortems on the Hitler War, namely *The Fatal Decisions* (1956, edited by William Richardson and Seymour Freidin) and *Decisive Battles of World War II* (1965, edited by Hans Adolf Jacobsen). The most recent exercise in the genre known to the author is John Gooch (ed.), *Decisive Campaigns of the Second World War* (special issue of the *Journal of Strategic Studies*, March 1990).

Creasy at least selected his battles on the criterion of their utility in contributing to the development of civilization, but too many of his successors have eschewed any definition of 'decisiveness', apparently on the assumption that dramatic bloodletting on a vast scale must have decided something.[6] The word 'decisive' is repeated like a mantra: seductive but all too often lacking precision or conviction. John Keegan focused on the point that the catch-all notion of 'decisiveness' has frequently diverted historians from carefully examining the reality of combat, that is, 'the face of battle'. My point here, taking up the earlier discussion of Napoleon's failure to convert decisive battlefield victories into enduring peace settlements, is that at least two further considerations need to be added to success on the battlefield: namely firm, realistic statecraft with specific aims, and the willingness of the vanquished to accept the verdict of battle. Ironically, decision through battle was becoming harder to achieve at the very time that Creasy's formula was popularized. In purely military terms, for example, Gettysburg appeared to be a decisive Union victory in 1863, but the Confederacy fought on for a further two years. The contrasting character and outcome of the two principal wars of German unification graphically illustrate a critical stage in this process.

The Austro-Prussian War in June and July 1866, also known as 'the Seven Weeks War', provides a rare example in modern times of a conflict between two great powers virtually decided by a single battle, including the preliminary engagements fought by the Prussian armies in their approach to Sadowa-Königgrätz. In retrospect this seems to have been an unnecessary war fought for inadequate reasons.[7] In A. J. P. Taylor's opinion, the chief issue at stake was Venetia. Austria promised to cede the province to France in return for the latter's neutrality, but she would lose it to Italy anyway in event of defeat. If victorious, Austria hoped to gain Silesia, whereas defeat was likely to mean exclusion from Germany north of the river Main. Napoleon III expected to gain Venetia, which he would hand over to his client Italy, and also to benefit from the creation of a new, independent German state on the Rhine which would serve as a barrier against a Prussian attack. Both sides horse-traded to secure benevolent French neutrality without knowing if the deal would be honoured by Napoleon. Even so, it proved difficult for Bismarck to engineer a war over Austro-Prussian administration of the duchies of Schleswig-Holstein taken from Denmark in 1864. Indeed, in A. J. P. Taylor's phrase, 'there was some technical difficulty in getting the war started', but Prussia resolved the problem by invading Saxony on 15 June.[8] There was no formal declaration—instead a trumpeter accompanied the leading unit to herald the start of hostilities.

In the light of Prussia's seemingly easy victories in Germany and Bohemia it is surprising that many foreign military commentators expected Austria to win, particularly as she dispatched her better army, commanded by the Archduke Albrecht, to Italy where it won a crushing victory at Custozza. But

Austria's recent record was more impressive than Prussia's, her long-service regulars were regarded as far superior to Prussia's conscripts, and the fact that the King of Prussia proposed to lead his forces in the field was held to be a fatal handicap. Confidently predicting an Austrian victory, Friedrich Engels wrote in the *Manchester Guardian* on 20 June, 'Prussia has had no great war for fifty years. Her army is . . . a peace army, with the pedantry and martinetism inherent in all peace armies.'[9]

Helmuth von Moltke, the chief of the Prussian general staff, carried out a much-criticized deployment by distributing his two main armies, containing some 300,000 troops, over an area of about 270 miles along the frontiers of Saxony and Bohemia. He had to reckon not only with an Austrian army approximately as large as his own, but also with the hostility of Saxony, Bavaria, Hanover, and several other German states. Moreover, since the King of Prussia insisted that Austria must appear the aggressor, the latter's commander, Ludwig von Benedek, was given the opportunity to make a speedy advance through the Bohemian mountains into Silesia. Benedek missed his chance, but the wide Prussian deployment entailed a gap of 40 miles between their First and Second Armies as they passed through the defiles of the Riesengebirge. The aged Jomini, and his numerous followers, including Willisen and Rüstow, condemned Moltke's wide dispersion and approach on exterior lines as strategic heresy and even the decisive outcome did not silence them.[10] But as his *Correspondence* reveals, Moltke had considered the variables of march timetables, distances, and enemy capabilities very carefully. He could only transport such large forces to the frontier by employing all five available railway lines (to Benedek's single line) and even then encountered serious supply problems as rolling stock accumulated at the railheads. Furthermore, he calculated correctly that Benedek lacked the offensive operational genius of Robert E. Lee and would not take risks to attack one of his armies in isolation. Indeed, as Jean Colin would later demonstrate,[11] Moltke's strategy so far from being reckless, was un-Napoleonic in its caution: he intended to unite his three armies (the army of the Elbe coming up on the Prussian right flank after taking Dresden) well *before* the battle and without attempting a turning movement to cut off the Austrians' retreat across the Elbe. There were only about 20 miles between Prussian First and Second Armies on 30 June and half that distance on the eve of battle (2 July). The enormous concentration of about 300,000 troops was crammed together on a front of some 20 miles.[12]

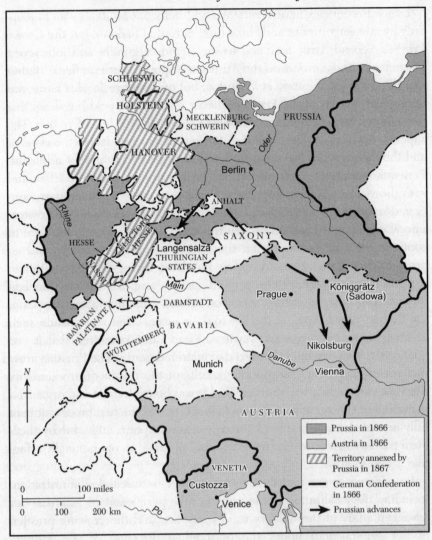

Map 4. The Austro-Prussian War, 1866

Prince Frederick Charles (First Army) briefly put Moltke's plan in jeopardy by attacking prematurely on a 9-mile front at Sadowa, but the Crown Prince's Second Army marched to the sound of gunfire and intervened piecemeal without orders on the Austrians' exposed northern flank. Rather like Blücher's intervention at Waterloo, but on larger scale, this move was decisive. A quarter of the Prussians never reached the field but those that did inflicted far higher casualties—some 40,000 for the loss of 10,000. The superior Austrian artillery kept the attackers at bay until late in the evening and the victors were too exhausted and disorganized to mount a pursuit. Benedek managed to extricate some 180,000 troops across the Elbe, and with the victorious army from Italy available to move to his support, continued resistance was expected. In fact this single day's battle—the greatest since Leipzig in 1813—had broken the Austrian army. Hostilities were to continue for another month but there were few clashes of arms and no prospect that Benedek would reverse the demoralizing defeat on 3 July.[13]

Contemporary military critics, particularly those like Engels who had predicted an Austrian victory, seized on Prussia's superior infantry weapon, the Dreyse 'needle gun', as the key to their success. But wiser heads, such as Dragomirov (who was an eyewitness) and Colin perceived that it was quality of the men behind the guns that mattered most.[14] The Prussian army was already displaying the positive results of the reforms introduced by Albrecht von Roon a few years earlier. It was better trained and organized than the enemy and its morale was high. Here was the first proof—not yet fully accepted—that universal three-year conscription, followed by thorough training in a four-year reserve, could produce an operationally effective nation in arms.

Prussia's other great advantages lay in organization and administration, including the localization of the army in nine permanent corps areas, her advanced study of the military use of railways, and the growing prestige of her general staff under the outstanding direction of von Moltke. Overconcentration on tactical analysis prevented orthodox military opinion from perceiving that, so far from perpetrating a terrible blunder, Moltke's bold and calculated strategic dispersal had been instrumental in preparing the way for the destruction of the enemy in battle. The most general lesson of the 1866 campaign, to be underlined four years later, was that in the new industrial age wars would be won by those nations which could raise, train, deploy, and command large conscript armies most effectively.[15]

It remains to answer the vital question: why was the outcome of a single battle accepted as decisive by both sides? As mentioned earlier, the Austrians had the military means to continue fighting. But Venetia was already forfeited and Silesia beyond their grasp. To continue the struggle would probably have entailed the enemy occupation of Vienna and, the perennial nightmare, the disintegration of the multinational imperial army. The memories of the revolutions of 1848 were too fresh to risk further resistance.

On the victor's side the quick and moderate conclusion of hostilities after Königgrätz was due to Bismarck's successful struggle to overcome the bellicose ambitions of the generals and the king. It must be borne in mind that no clear and agreed structure for civil–military relations was in place. Moltke was only just, in the course of the war, establishing his authority as the King's chief military adviser and the fount of all orders to the army commanders. In principle he accepted the primacy of political authority in peacetime but emphatically not during military operations. Bismarck, whose own precise position and powers in the chain of command were also uncertain, antagonized Moltke from the outset by issuing orders directly to field commanders without consulting the chief of the general staff.[16]

Tension between them mounted rapidly in the days after Königgrätz, Moltke was determined to press on relentlessly to destroy the remaining Austrian forces, while the King wished to crown the victory by a triumphal entry into Vienna. Both wanted a total victory and a punitive peace. At Prussian headquarters there was talk of the need to acquire Saxony as well as parts of Bavaria and Bohemia.

Bismarck, by contrast, was concerned from the outset to treat Austria moderately since he was acutely aware of the potential dangers raised by Napoleon III's diplomatic intervention, only two days after the victory, to propose an armistice and the beginning of peace talks. As Bismarck wrote to his wife on 9 July:

If we are not excessive in our demands and do not believe that we have conquered the world, we will attain a peace which is worth our effort . . . I have the thankless task . . . of making it clear that we don't live alone in Europe but with three other Powers who hate and envy us.[17]

In the following days Bismarck intervened boldly to prevent the generals pressing home their advance to Vienna and beyond; at one point asking sarcastically why it was necessary to stop at the enemy capital? Why not pursue the Austrians into Hungary, go on to Constantinople, found a new Byzantium, and leave Prussia to her fate? This was a veiled reference to a possible French attack in the Rhineland while Prussian forces were entangled deep in the Austrian Empire.

The civil–military crisis came to a head on 19 July. Bismarck decided to accept French proposals for a peace settlement which would halt operations just as the Prussians were poised to cross the Danube and smash the remaining Austrian resistance. After a long and stormy meeting at the royal headquarters, Bismarck got his way, no doubt helped by von Roon's disquiet about the Danube crossing on military grounds. Napoleon III justified Bismarck's gamble in accepting a ceasefire by subsequently raising no objections to substantial Prussian territorial gains in northern Germany, including Schleswig-Holstein, the kingdom of Hanover, the duchy of Nassau, and the free city of Frankfurt. These annexations involved the deposition of a king, an electoral prince, and two dukes.

Even these rich pickings, which added some seven million citizens to Prussia, did not satisfy the King who thought that Saxony, Austria's ally, should not be spared, and that Austria herself should forfeit territory and pay an indemnity. He feared that the generals would reproach him for not obtaining adequate rewards for their victory. Only after long emotional arguments did Bismarck persuade the King to sign the peace preliminaries on 26 July and then he did so with bad grace.[18] Moltke and the senior commanders made no objections to the peace terms but made it clear to Bismarck that they bitterly resented his interference in professional military matters and would ensure that he would not do so again in any future war.

The eventual peace treaty, signed at Prague on 23 August 1866, confirmed Prussia's predominance in the new North German Confederation, from which Austria was excluded, but the latter was spared any deep humiliation. She lost no territory, paid only a small indemnity, and suffered no triumphal march through Vienna or military occupation. Prussia, in addition to her territorial gains in Germany, levied indemnities on the German States which had fought against her, and signed separate treaties which in wartime would give her control of the armies and railways of Bavaria, Württemberg, and Baden.[19]

The victory at Königgrätz and its political results cleared the way for Prussia's domination of Germany. Without the dissolution of the German

Confederation and the territorial annexations this would not have been possible. France had also suffered a diplomatic defeat because Napoleon had miscalculated the extent of Prussia's accretion of power. In effect the balance between Austria and Prussia had been destroyed and henceforth the former would depend increasingly on the support of Prussia/Germany for her Balkan ambitions. Last, but not least, the spectacular military success 'cast a magic spell' over Prussian domestic politics, reducing liberal opposition to Bismarck's military and foreign policies.[20]

Enough has been written here to show that the Prussian campaign in 1866 was a brilliant example of war as an instrument of state policy. As the battle casualties indicate, it was conducted with ferocious, all-out commitment but for specific limited political objectives. The generals were able to deliver the prize of victory in a decisive battle completed in a single day. The generals' further ambitions, which would have prevented a moderate peace were restrained, albeit with difficulty. For all the passions generated and the heavy casualties it remained a 'Cabinet War' in the old style. Finally, far-reaching territorial and constitutional changes, also reminiscent of an earlier age, were accepted by those directly affected—and by the great powers—without violent objections.

Meanwhile the American Civil War had displayed some contrasting features which, in a longer perspective, have caused it to be described as 'the first modern war'. The preservation or fragmentation of the Union provided a fundamental constitutional issue, but one which Lincoln's government initially hoped to resolve by a limited military effort which would culminate in the occupation of the city of Richmond. However, the announcement of the preliminary Emancipation Proclamation after the battle of Antietam in October 1862 made the North's aim more 'total' and punitive, thereby extending the object of the war to an unlimited extent. The conflict thus began with Jominian ideas in the ascendancy, focusing on lines of operation and occupation of territory, but developed into a Clausewitzian 'total war', encapsulated in Sherman's remarks 'War is cruelty, and you cannot refine it' and 'We are not only fighting hostile armies, but a hostile people, and we must make old and young, rich and poor, feel the hard hand of war'.[21] The transition from limited to near-total war was also evident in the initial false assumptions that the conflict would be short, that only limited numbers of soldiers would be needed, and that a decisive battle would end the rebellion. The vastness of the country with its sparse population and tenuous communications, quite apart from the complexity of political allegiances, meant that the Confederacy's 'centre of gravity' was difficult to locate. Neither side was able to occupy the other's capital city, and even such a coup might not have ended the war. European intervention on behalf of the Confederacy could well have done so, but this hope faded after 1862.

Perhaps the sharpest contrast to the contemporary Prussian experience was that, although Confederate generalship secured several brilliant victories, usually against superior numbers, in 1861 and 1862, none was decisive. Nor, when the Union forces began to gain the ascendancy in 1863–4, were their victories, such as Gettysburg and Vicksburg, any more decisive, at least in the short term. Though great battles were highly relevant to the course of the war, they were only one factor in the attritional process as the Confederacy was drained of manpower, blockaded, divided, and subdivided by the Union advances, and eventually overrun. More acutely than in Prussia's wars (because popular opinion was so closely involved), the Civil War exposed the enormous problems in establishing civilian, political control of operations and objectives. Lincoln succeeded in steering the North to a clear-cut victory, as a consequence of which the Union was preserved and slavery abolished. But the costs were great and the wounds slow to heal. No military theorist was likely to take this as a model of 'decisive victory'.

The American Civil War was indeed observed by numerous European journalists and professional military men, and some of its features were intensely and even obsessively written about and studied in Europe right up to, and even after, the First World War.[22] But most contemporary students of the war were interested in a specific aspect, such as infantry or artillery tactics; they were not concerned to place the conflict in the broad context of political, technological, and social development. In any case the war had so many unusual aspects—not least that it was a civil war—that it was all too easy to write it off as exceptional. Indeed, so dominant was the Jominian tradition by this time that professional attention focused narrowly on generalship, battles, and tactics.[23] In sum, it was virtually impossible for European military men to perceive, or admit, that the long, attritional Civil War, rather than Moltke's decisive victories, provided the more reliable guide to the evolving character of modern war.

The completeness of Prussia/Germany's victory over France in 1870–1 was even more startling to contemporary pundits than her defeat of Austria. National pride, fostered by the Napoleonic tradition, remained strong, and

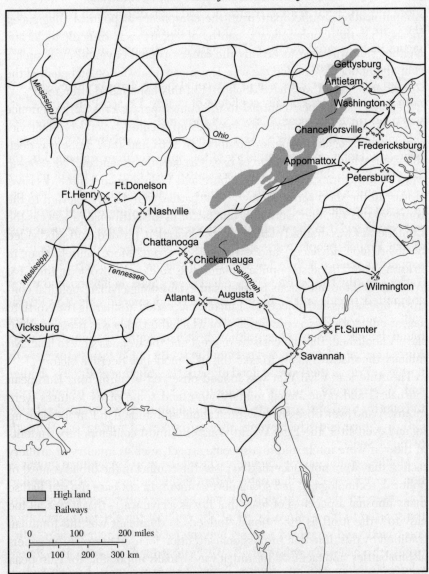

Map 5. The American Civil War, 1861–1865

more recently French military prowess had been displayed in North Africa, the Crimea, and Northern Italy. Also, too much attention on weapons and tactics had obscured the overall strength and efficiency of the whole Prussian military system.

This was another case where war could hardly have occurred had not both parties been reasonably confident of victory. Franco-Prussian hostility was deep-rooted in recent history, yet the causes of war in 1870 were even less substantial than in 1866. As a consequence of a relatively unimportant issue (the Hohenzollern candidacy to the Spanish throne enflamed by the Ems telegram), the French and Germans seemed bent on a test of national honour and military strength.[24] Bismarck's later claim that he had engineered the whole crisis is doubtful, but he certainly improvised adroitly to put Napoleon III in the wrong when the opportunity arose. The great mass of the French people were apathetic, but Napoleon and his imperialist supporters wanted a showdown: a case of war for its own sake. In A. J. P. Taylor's words 'The Second Empire had always lived on illusion; and it now committed suicide in the illusion that it could somehow destroy Prussia without serious effort'.[25] Contrary to her hopes and expectations, France found herself without allies; although still resentful of defeat in 1866, Austria remained neutral, and even in the southern German states recently hostile to Prussia there was 'a flood of patriotic exaltation'.[26] Bavaria, Baden, and Württemberg mobilized promptly and efficiently to produce contingents for the Third Army, which had the honour to be commanded by the Crown Prince and which would play an important role in the climactic battle of Sedan.

Moltke's contingency planning produced the most streamlined mobilization ever witnessed to that date. Within eighteen days 1,830,000 regulars and reservists passed through German barracks, of whom 462,000 were transported to the western frontier.[27] By contrast the French mobilization was chaotic and produced scarcely half the total of German soldiers. After twenty-three days French reservists were still struggling into their regimental depots, often without essential items of kit or weapons. French mobilization in short resembled a disturbed ants' nest. Against a well-armed and numerically superior opponent with a more efficient supply system, an experienced and confident high command and general staff, and a thoroughly rehearsed war plan, the French tradition of 'muddling through' (*On se débrouillera toujours*) was cruelly exposed. While it would be too deterministic to say that France had already lost the war before shots were fired, she began operations under a handicap from which only united and inspired leadership at the top could rescue her.[28]

Far from this hope, the French army suffered from the disastrous higher direction of the war by Napoleon III, compounded by the machinations of his empress and the irresolution of his principal army commanders Bazaine and Macmahon.[29] In the opening battles at Spicheren and Worth the German casualties were heavier but they kept on attacking, supporting dispersed units by marching to the sound of the guns. French valour was undermined by continuous retreat and the absence of clear orders. Moltke's dictum that operational plans never survived the first contact with the enemy was vividly demonstrated in his own experience. Completely misreading the movements of the sluggish Bazaine, whom he was supposedly pursuing westward from Metz, Moltke crossed the Moselle, only to discover that he had got behind the French and cut their communications— a classic Napoleonic move achieved by accident. A series of desperate, closely contended actions culminating in the battles of Gravelotte-St Privat on 18 August caused Bazaine to withdraw into the fortified area of Metz. MacMahon's army, with the desperately sick Emperor as its nominal head, set out to relieve him but was pinned against the Belgian frontier at Sedan and pounded by longer range artillery which the French could not even engage. On 1 September Napoleon surrendered and went into exile, while more than 100,000 French troops passed into captivity. This disaster caused a revolution in Paris; a Government of National Defence was set up and the Third Republic proclaimed. France's last hopes of retrieving the situation by conventional warfare now lay with Bazaine, but after a lacklustre conduct of the siege, he surrendered Metz at the end of October.

These two dramatic defeats, involving the destruction of the bulk of France's regular soldiers and the organization of her two principal armies, which precipitated the flight of Napoleon and the end of his empire, were as 'decisive' as could be imagined. In traditional terms these events should have resulted in peace negotiations. Bismarck had indeed placed considerable hope in keeping Bazaine's army in being because he would have preferred to negotiate with the Emperor as representing a stable, conservative regime.[30]

The new Republican Government of National Defence in Paris had other

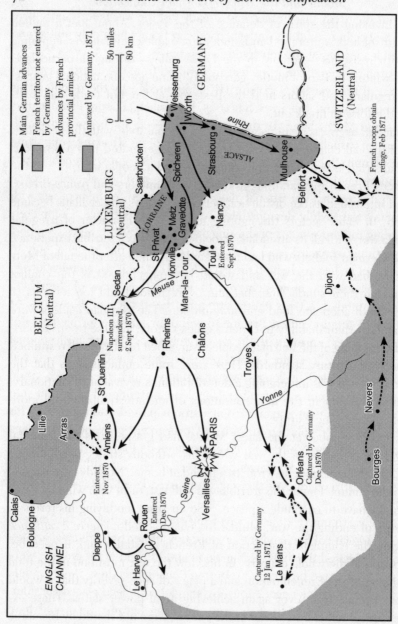

Map 6. The Franco-Prussian War, 1870–1871

ideas, invoking the spirit of defiance of the 1790s. About half a million amateur soldiers were raised and, under some able commanders, kept up a stubborn resistance—notably in the Loire valley—through a particularly severe winter. In Paris a motley garrison of some 300,000 national guardsmen, regulars, and sailors held the besiegers at bay but made only half-hearted efforts to break out.

What had begun as a campaign in the traditional style, with the decencies of professional conduct generally observed, degenerated into a ferocious unequal struggle between an invading army and a people in arms.[31] In the provinces French regulars fought alongside volunteers and *francs-tireurs*. Hatred intensified as the distinction between soldiers and civilians became blurred. In many sectors the fighting took on the character of irregular guerrilla war, with all its attendant atrocities. This was a terrible experience for the German soldiers who had mistakenly assumed after the fall of Metz that they had won the war and would be going home. As the conflict dragged on their supply lines became overextended and too tenuous to provide both adequate food and ammunition. The German leaders were frustrated by their inability to clinch a quick victory, and the troops experienced something of the hell of a people's war, albeit brief and still restrained by twentieth-century standards. Here were some ominous signs that the character of war was undergoing a deterioration: a revival of revolutionary, 'absolute' war but in the new conditions of an industrializing and more fiercely nationalistic Europe.

By mid-December at the latest Moltke had grasped this fundamental change in the nature of the war which his orthodox strategy had already 'won'. On 18 December he wrote to a friend at home: 'Nobody here has any idea of how much longer this terrible war will last, or of with whom we will eventually have to conclude a peace.'[32] So far from modifying his terms in the hope of ending the war, Moltke too championed a *guerre à outrance* involving the complete destruction of French resistance and, if need be, occupation of the whole country. In his view the war must end in the total submission of the French nation and a peace of such severity that it would prevent France from ever again achieving great-power status. These extreme views, fuelled by a hatred of the enemy, were not shared by Bismarck and were bound to intensify the friction in civil–military relations which had been apparent since the very outset of hostilities.

Whereas the War Minister, von Roon, was reluctantly admitted to the daily conferences with the King at the field headquarters, the Chancellor was pointedly excluded and not even kept informed of the progress of operations or of future military plans.[33] Bismarck accepted this position initially, but for six weeks after Sedan, when even small-scale military movements might have some potential bearing on peace negotiations, the cold-shouldering continued. Not until 15 October were arrangements made to inform him of messages sent to the German press, but even this concession was operated in a slipshod manner.

From an early stage in the war the Chancellor was anxious lest foreign diplomatic intervention, particularly Russian, might end hostilities prematurely and deprive Germany of her political objectives. He was also acutely aware of the need for a stable French government capable of signing and implementing a peace treaty. Hence his vain attempt to preserve Bazaine's army while conducting delicate negotiations with Napoleon, which was frustrated partly by Prince Frederick Charles's deliberate obstruction of the imperial go-between, General Bourbaki. Moltke shared the Prince's opinion that the capitulation of Metz was a purely military matter outside Bismarck's remit. The issue epitomized the chief of staff's adamant position that a clear distinction could be drawn between military and political responsibilities. This and similar disputes had already soured relations between the chief of the general staff and the Chancellor when, towards the end of November, the question of the siege of Paris brought the contretemps to a head.[34]

Moltke and his senior colleagues marshalled the conventional military arguments against bombarding a great city like Paris with its surrounding girdle of fortresses. Recent experience, such as the allied siege of Sebastopol, was held to demonstrate the unwisdom of the project: it would be very difficult to assemble sufficient guns and ammunition; storming the defences would be extremely expensive and destructive; and possible failure would be bad for morale. Bismarck brushed these objections aside: it was politically imperative to capture Paris and end the war before an international conference did so over his head. Russia's recent abrogation of the Black Sea restrictions placed on her in 1856 made such a conference likely. Moltke gave way and the bombardment began on 27 December, but he deeply resented the pressure put on him, particularly through the Chancellor's manipulation of the German press. An attempt by the Crown Prince to improve relations between the two men with a dinner party only served to expose the width of the gulf between them.[35]

The real issue was that, whereas Bismarck saw the capture of Paris as a means to end the war on moderate terms, for Moltke it would merely be the prelude to a war of extermination against all outstanding French resistance. This was made clear in a memorandum which Moltke addressed to the King in mid-January. He insisted that the city of Paris as well as the forts should be occupied. The city should be administered by a German governor, regular troops and *gardes mobiles* must be disarmed and sent as prisoners to Germany, and all the enemy's military standards handed over. Bismarck's war aims against France were also substantial but he wished to spare the capital any unnecessary humiliation. The King had previously seemed to be more in sympathy with Moltke and the high command, but on 20 January he changed his position by authorizing Bismarck to begin armistice talks. Five days later two royal orders confirmed that henceforth the Chancellor was to be kept fully informed about military operations. Moltke was so shocked that he talked of resignation. Instead, he sent a memorandum to the King showing that he continued to regard his own office and the Chancellor's as equal and mutually independent agencies under the royal command. He only conceded that they should keep each other fully informed. As Gordon Craig comments, this was a remarkable assertion of rights, considering that as recently as 1859 the chief of the general staff had not been permitted to report directly even to the War Minister.[36] The King did not reply and it was Bismarck who successfully conducted the armistice negotiations from 26 January onwards. The siege of Paris ended two days later.

Moltke and the other generals at field headquarters openly expressed their detestation of the Chancellor for his interference in professional military matters. The chief of the general staff continued to inculcate his opinion that politics and strategy must be sharply differentiated in wartime: when the guns speak the politician must fall silent until the commander has delivered the victory.

Although Bismarck's peace terms left France undisputedly a great power, they were harsher than those negotiated with Austria, which he regarded as a natural ally. The German generals had no cause for complaint that the Chancellor had been too lenient, though whether they had put pressure on him or not is disputed. France was to support a German army of occupation until she had paid a war indemnity of 5 billion francs (which in the event only took three years). France was allowed to retain the important fortress of Belfort, but only on condition that she agreed to a triumphal German march through Paris. More controversially, she was forced to cede the strategically and industrially important province of Alsace and one-third of Lorraine.[37] Bismarck had been bent on this annexation from the outset, on the assumption that the French would not forgive or forget their humiliation and might attempt revenge. It was therefore prudent to secure a buffer zone to deter such an operation. Of course Bismarck still hoped that the French would become reconciled to the loss of these provinces, but as A. J. P. Taylor remarked, 'A hereditary monarch can lose provinces; a people cannot so easily abandon its natural territory'. Alsace and Lorraine were not essential to Germany's predominance in Central Europe, but they *were* a symbol of France's lost greatness which even their recovery in 1919 failed to restore.[38]

The greatest gain from the victory over France was that it made possible the creation of a united German Reich with William I as its (initially reluctant) Emperor. In the euphoria of victory, in which their contingents had played an important role, it was virtually impossible for the political leaders and rulers of the south German states to stand aloof from incorporation in the Reich. The price of independence would be political isolation, economic decline, and unbearably high defence costs. In any case the Chancellor displayed great delicacy in dealing with the domestic prerogatives and ceremonial rights of Bavaria, Württemberg, and Saxony, while ensuring that their armed forces would pass under Prussian control in event of war.[39]

Prussia/Germany's outstanding success in the war of 1870–1 illustrates the ambiguity of all military victories, however clear-cut. Moltke, who in his attitude to political control, his quest for total victory, and belief in the inevitability of war, represented 'the military mind', was soon expressing regret that France had not been more completely crushed in 1871.[40] This was surely a delusion. France's rapid recovery and rearmament in the 1870s and Russia's belated awakening to the danger posed by a united Germany led to an arms race of unprecedented intensity and duration. As early as 1875 and again in 1877 Moltke spoke of the desirability of a preventive war against France, and as late as 1887 he recommended a forestalling action against Russia. All these proposals were rejected by Bismarck, whose decisions were calmly accepted by Moltke. The latter seems to have become increasingly pessimistic about the prospects of carrying out such a risky operation.[41]

This uncertainty and growing pessimism was evident in Moltke's contingency planning, which always assumed Russo-French co-operation to confront Germany with a simultaneous war on two fronts. In an elaborate plan worked out early in 1877 Moltke argued that because France could mobilize more quickly than Russia, a preventive attack would have to be launched on the former as soon as there was a real threat of war from the latter. He calculated that France could be defeated in three weeks, but there would be insufficient time to exploit the victory because eight army corps would have to be transferred to the eastern front to meet the Russian offensive. But what would happen if, even after a 'decisive' battle, France was not prepared to make peace? The experience of the winter war in France in 1870–1 suggested to Moltke that this was all too likely, but he had no solution to offer. He could only admit that the successful outcome of such a preventive war could not be guaranteed.[42]

In 1879 and the following years Moltke worked out a plan for an offensive in the East designed to encircle and surround the Russian armies in Poland. A surprise, preventive attack should have made this feasible, but Moltke was too realistic to believe that such a victory could be exploited by an advance deep into Russia. Here again the diplomats would have to negotiate peace terms on the basis of a single big victory so that the bulk of the forces could be switched to the other front. Moltke was drawn increasingly by these calculations to the grim conclusion that in a two-front war it would be impossible to knock out one major enemy by a decisive battle in a cabinet war of the old style. As he stated in a famous Reichstag speech on 14 May 1890: 'The age of cabinet wars is behind us—all we have now is people's war, and any prudent government will hesitate to bring about a war of this nature with all its incalculable consequences.' He sketched a terrible attritional war between the great powers, not one of whom would accept the verdict of battle and the ensuing high peace terms, and concluded: 'Gentlemen, it maybe a war of seven years' or of thirty years' duration—and woe to him who sets Europe alight, who first puts the fuse to the powder keg!'[43] In his final years Moltke even suggested that the main role of armed forces should be to act as deterrents against the danger of uncontrollable people's war. Here was an appropriate doctrine to preserve the conservative social order of Imperial Germany and the territorial boundaries established by Prussia's earlier victories.

This was not, however, the prevailing outlook in the German general staff after Moltke's retirement. Instead, his successors were torn between the other options of deliberately waging all-out people's war or finding a formula for a short, decisive campaign. It was the latter option, which one historian has described as 'trying to square the circle', which attracted Schlieffen and his successor, the younger Moltke. In other words, given her probable inability to sustain a long war against heavy numerical odds, Germany's best chance lay in developing her army as a rapier rather than a bludgeon, in the hope that superior quality would prevail in a short and 'limited' war like those of 1866 and 1870.[44] This topic will be pursued in the next chapter. Here it has been suggested that the campaigns of 1866 and 1870–1 come as close as can be imagined to the Clausewitzian formulation of war as an instrument of state policy. They were comparatively short and inexpensive in casualties; they were popular at home; their conduct did not escape from political control; and they produced substantial territorial and other advantages for the victor.

A sombre lesson, which the elder Moltke appeared to accept towards the end of his career, was that changing conditions in social and political life, as well as within war itself, might make such decisive Cabinet Wars unrepeatable, but this was not the practical deduction of the European military (and naval) professionals. At the operational level this was not at all surprising. Officers needed to study the techniques of their profession which would enable them to defeat the enemy in war. It was the responsibility at the highest level of generalship and statesmanship to address the gap between the operational level and grand strategy. At least this should have led to the appreciation that, even if the commanders pulled off a decisive battle in military terms, it might not be decisive politically.

Moltke's acknowledgement of Clausewitz's influence helped to enshrine the Prussian theorist as the chief inspiration and presiding genius of the wars of 1866 and 1870.[45] His emphasis on the untrammelled drama of war, and the vital importance of morale, leadership, and patriotic fervour to overcome uncertainties and inertia, were now more appreciated. In more practical terms his stress on the importance of mobilizing maximum numbers and employing simple direct methods to grip the enemy in a decisive battle had been vindicated. The Prussian short service conscription and reserve system had proved its value, setting new standards both in quality and numbers that other European states strove to emulate. Indeed, military power was increasingly assessed by the number of troops who could be mobilized in a given period and the echelons of trained manpower held in reserve. Railways, for all their limitations, were regarded as the critical factor in contingency planning for war. Thanks to Moltke's towering achievement, general staffs were becoming recognized as the essential 'brain of an army': their authority in both planning and operational command was growing. Finally, the last quarter of the nineteenth century witnessed rapid technical innovation and fierce national competitiveness in offensive weapons, communications, and defences. In the fast-changing political and military environment no serious student of war thought it would be easy to emulate the decisive victories of 1866 and 1870. The disturbing question was really whether it could be done at all.

The Quest for Victory in the Schlieffen Era
1890–1914

Historians are very properly reluctant to use the word 'inevitable', but from about 1890 onwards public figures displayed an almost uncanny sense of an impending great war between the European powers. This was surely the last period when war could be discussed not simply in terms of catastrophe, but also positively and even ecstatically as a social and biological necessity, even a tonic. For serious students of the phenomenon, the prevailing model was that bequeathed by the elder Moltke: total victory in a short decisive campaign.

From the vantage point of post-1918 we are inclined to scan pre-1914 pronouncements about the nature of future war with a very critical eye, alert to spot naïvety, errors, and ironies. But while some foolish statements were certainly made, for example by the then Colonel Foch in asserting that every improvement in fire-power would be to the advantage of the attacker, by and large the professional experts were acutely aware of the difficulties of securing Napoleonic-style decisive victories on the modern battlefield. Since great conflicts would surely occur, the professionals' task was to consider how to win them: they could not afford to appear defeatist. In any case the situation did not seem hopeless. The enormous growth of armies, their means of transport and communications, and their destructive fire-power all seemed ominous; but lesser contemporary conflicts were anxiously studied for more positive, reassuring signs that the Napoleonic-Moltke model was still relevant and repeatable in changed conditions.

One of the areas in war planning and preparations where the professionals are generally held to have gone most badly astray is in the emergence after 1900 of both tactical and strategic doctrines inculcating the offensive. Historians and political scientists alike have made merry in criticizing these concepts in the light of the marked superiority of the defensive in most theatres and for most of the time in the First World War.[1] Some historians, such as Douglas Porch and Michael Howard, have recently shown some sympathy for, and understanding of, the dilemma of pre-1914 theorists, stressing for example their awareness of the lethal fire zone created by the rapid development of fire-power. An Israeli scholar, Gideon Akavia, approaching the issue from a non-historical background, has put the matter in a new and more convincing perspective.[2] Rather than focus obsessively on the supposed 'cult of the offensive', it makes more sense, he argues, to give primacy to 'the cult of the decisive victory'. In other words, when the military leaders were faced with the political requirement for a decisive victory, they had no choice but to adopt a doctrine whose military-technical character had many offensive features. If any war could produce the very ambitious political goals demanded, it had to be fought offensively. By concentrating on political requirements more than on military doctrine, Akavia draws a different moral from the critics of the 'offensive school': namely, that military means in general are a dangerous tool for providing security, even when necessary. Here we may add the reflection that, for a variety of reasons to be explored in this chapter, the military instrument of 1914 was ill-suited to produce the quick and clear-cut victory demanded of it in the prevailing conditions.

A further word of caution is necessary. Historians may legitimately enjoy the benefits of hindsight, but they must constantly remind themselves that the individuals they criticize could not know in detail how events would develop. This banal reflection applies particularly to watersheds like 1914 which can easily be made to appear the natural and inevitable outcome of everything that happened in the preceding decade. True, from the mid-1890s the rival alliances (Germany, Austria-Hungary, and Italy versus France and Russia) seemed the most likely belligerents in the impending war, but the final alignments were far from clear until 1911, and even after that there were many unknown factors. Where would the war start, would it immediately become a general European war, and would peripheral states such as Italy, Britain, and the Ottoman Empire be involved? How far was it possible to predict in advance the general military, political, and economic character of future war without knowing which were the belligerent powers, where they would fight, and which laws and customs of war they would honour or defy?

Although there were periodic 'war scares' and, in the case of Britain, 'invasion scares' in the 1880s and 1890s, it was really after 1900 that the dangerous notion of preventive war took root, especially among the general staffs of the great powers. Wolfgang Mommsen has explored the complex development of this thinking in Germany where the military felt themselves to be under increasing domestic pressure, while at the same time the nation's defences and foreign policy were in confusion.[3] These fears came to a head at the time of the Agadir crisis in 1911 when August Bebel captured popular anxieties in writing: 'So there will be armament and rearmament on all sides until one day one or other side will say: rather an end in horror than horror without end . . . If we wait longer we are the weaker side instead of the stronger.'[4]

Schlieffen and the younger Moltke had considered the possibilities of preventive war during the first Moroccan crisis of 1905 and subsequently, without necessarily advocating it, but from 1911 onwards the tone of the military planners and polemicists became more fatalistic and more urgent. General Friedrich von Bernhardi, for example, in his notorious book *Germany and the Next War* (1912) bluntly remarked that 'there is no way in which we can avoid going to war for the sake of our position as a world power and that we should not be concerned with postponing it for as long as possible, but rather should concentrate on bringing it about under the most favourable conditions possible'.[5] At first, mainly through the right-wing nationalist press, but eventually through bourgeois, Centre Party newspapers as well. German public opinion became deeply influenced by the notion of inevitable war. In the early 1900s popular enthusiasm had been directed mainly against Britain and, to a lesser extent, France, but by 1913 German fears were focused increasingly on Tsarist Russia. France, of course, was still involved indirectly as the main financier of Russian rearmament. By the spring of 1914 there was a widespread assumption among decision-makers in Germany, shared by Moltke and the Kaiser, that Russia would launch a preventive war in 1916 or 1917 as soon as her rearmament and strategic railway developments were completed. This belief caused profound anxiety because it undermined the fundamental assumption in the Schlieffen Plan that Russia would mobilize slowly and would not be able to mount an offensive until after France had been defeated. Consequently the idea gained ground, not only in military circles, but also amongst the public, that Germany should seize an opportunity to launch a preventive attack on Russia while her rearmament was still incomplete. These reckless ideas were not of course directly translated into action, but in the July 1914 crisis they contributed to the German leaders' fatalistic acceptance that they could not control the slide towards war. Twenty years of nationalist agitation and propaganda about the inevitability of a great European war had acquired the quality of a 'self-fulfilling prophecy'.[6]

There is no need here to discuss the numerous service leagues and pressure groups which did so much to foster militarism and nationalism before 1914, with their insistent emphasis on the impending great war which would settle the fate of Europe.[7] But it is worthwhile to illustrate prevalent ideas about the significance of war, its main features, and future prospects through the books of two best-selling militarist writers—Major (later Field Marshal) Colmar von der Goltz and former cavalry general Friedrich von Bernhardi—not that comparable authoritarian polemicists could not be found in other nations.

In *The Nation in Arms* (first published in 1883 but with a further five editions by 1900), von der Goltz exhorted the German people to prepare for further great military efforts. Striving for political and civilizing ideals was bound to result in war. 'True it is wars are the lot of mankind; are the inevitable destiny of nations. Eternal peace is not the lot of mortals in this world.'[8] Writing in 1883 he expressed doubt about 'the cleansing effects of the last war'. He also doubted whether the rapid victories of 1866 and 1870 could be repeated due to the ever-growing size of armies. 'Millions cannot be tossed hither and thither like thousands.' Nearly a decade before Schlieffen began to grapple with the problem, he noted that there was barely enough room on the whole Franco-German frontier for the two armies to deploy.[9] He therefore envisaged a protracted struggle with little scope for movement. The outcome would be the complete annihilation of one belligerent or the total exhaustion of both. Enhanced national consciousness would tend to prolong resistance. On the whole, he reflected, 'It has become much more difficult to coerce a great state into yielding than was formerly the case.' Furthermore, once national feelings have been aroused, the nation becomes *one and indivisible* so that *annihilation* must mean something quite different to what it did in Napoleonic times.[10] Three steps or stages might be necessary to obtain victory. First, the enemy must be deprived of any hope of victory by the destruction of his armies in the field. Next, one's adversary must be deprived of hopes of a change of fortune by the capture of his capital and the occupation of his strongholds and of districts where his forces might be reorganized. Meanwhile diplomacy will cut off all hopes of extraneous assistance. Lastly, pressure should be exercised on the most prosperous districts and, if necessary, the whole country be occupied, thus cutting off communications with the outside world. In conflicts with equal opponents the nation must be prepared to wage all-out war.[11] He concluded with an eloquent appeal to German citizens to fight for glory and immortality:

Whoever has a heart, feels it beat higher and becomes enthusiastic for the profession of a soldier . . . The star of the young Empire has only just risen on the horizon . . . And if ever a State afforded a guarantee of long existence, so it is a strong, united and military Germany in the midst of the Great Powers of Europe.

Inspired by a passionate love of the fatherland, the German nation in arms will be assured of victory in the coming conflict.[12] This emotive rhetoric evaded the issue of whether Germany had the manpower and resources to impose such draconian terms on one major opponent, let alone a hostile alliance.

Friedrich von Bernhardi gained renown and notoriety when he published *Germany and the Next War* in 1912, followed shortly afterwards by an even more bellicose effusion entitled *Our Future*. In the former book he contended that war was essential to promote the progress of mankind. He believed that those intellectual and moral factors which ensure superiority in war are also those which render possible all positive developments: 'Without war, inferior or decaying races would easily choke the growth of healthy budding elements and a universal decadence would follow.'[13] Early chapter headings included 'The Right to Make War', 'The Duty to Make War', 'Germany's Historical Mission', and—most revealingly—'World Power or Downfall'; yet he claimed that he had to attack 'the poison of peace' in order to prepare the pacific German nation for the inevitable struggle. He paints a nightmare scenario of Germany attacked on two or even three fronts by Britain, France, and, perhaps, Russia also. Germany will be heavily outnumbered and will not (he correctly predicted) be able to rely on Italian assistance. It would be a bitter struggle for existence because Germany's enemies would seek to annihilate her on land and at sea. German defeat would 'check the general progress of mankind in its healthy development, for which a flourishing Germany is the essential condition'. 'We must therefore prepare not only for a short war, but a protracted campaign.'[14]

Having outlined a realistic, albeit overdramatic, scenario of Germany on the defensive against a vastly superior alliance, Bernhardi, towards the end of his study, switches to the offensive strategy favoured by Schlieffen and his successor. Salvation lies in attacking and defeating one antagonist before dealing with the others. It is up to German diplomacy, he concludes, to provoke France to commit an act of aggression so that Germany can attack her in the expectation that Russia would for a time remain neutral.[15] Here, in broad terms, though not in detail, was a hint of the dilemma which Germany actually faced in 1914: the need to find some pretext to launch an offensive against France even if hostilities might first occur in the East.

Bernhardi's militaristic ideas as described here and his grandiose visions of German expansion in Central Europe, the Balkans, and, above all, in Africa were not surprisingly denounced in German government circles and the liberal press as irresponsible and harmful to foreign relations. But shorn of its colourful rhetoric, this polemic was in broad terms an expression of the policies which Germany would follow in 1914 and through the ensuing conflict.[16]

As I. F. Clarke's brilliant survey demonstrated,[17] a great deal can be learnt from fictional writing about future wars, a genre which enjoyed one of its most popular spells in the period between 1870 and 1914. Although some of the stories were futuristic and far-fetched, such as Sir George Chesney's trend-setting thriller *The Battle of Dorking* (1871), which postulated an invasion from a nation, Germany, which, as yet, had virtually no navy; nevertheless the genre as a whole conveyed an accurate sense of public anxieties about contemporary war threats and the likely forms which operations were expected to take. Harder to assess is the influence of imaginary war writing in shaping national stereotypes and affecting government policy.

In the 1880s the most frequently imagined European war in literature was between France and Britain. French writers, as in the anonymous *Plus d'Angleterre* (1887), appealed to chauvinistic prejudice by presenting an idealized picture of a united French nation humiliating the old enemy. Some accounts are so nationalistic and angry that they border on farce. Clarke makes the interesting observation that French anti-German fiction was less passionate and more realistic than the anti-British variety because it was mostly written by army officers. This touches on an important distinction between fiction whose ultimate purpose was to stimulate improvements in the armed forces and the national defences, and the more sensational variety designed to sell newspapers and books by playing on public fears of devastation, slaughter, rape, and enemy occupation.

From the early 1890s, following political developments, a Franco-German war of revenge—sometimes including France's new ally, Russia—became topical. A prolific French author who followed this trend was Captain Emile Danrit (pseudonym of Emile Driant) who, in *La Guerre en rase campagne*, lamented that the great war which all were expecting was still delayed. 'To pass the time whilst waiting, I have dreamed up this Holy War in which we will conquer, and this book is my dream.' This, and comparable fictions in German and English, represent 'perfect para-utopias for an epoch of exaggerated nationalism'. All express the desire for total victory over the loathsome, ancient enemy. In these unshaded fantasies the superior virtues and vigour of the fatherland overwhelm a despicable enemy lacking honesty, honour, and intelligence. Suddenly transferred from myths relating to a legendary past to the near future, and based on real national rivalries, these 'popular epics for a period of universal literacy' offered a seductive ultimate solution to national problems. Consequently, victory was always absolute, usually culminating in the triumphal entry into the enemy capital and his humiliating surrender. Thus Danrit concluded his anti-British story *La Guerre fatale* with the reading of peace terms in the ruins of the House of Commons. As Clarke comments, in this heyday of jingoism, typified by a new emphasis on national anthems, flags, magnificent uniforms, and ceremonies: 'the aggressive nation-states . . . had everything on their side except common sense. Most of all in the nineteenth century the nation-state meant the glories of war.' This outpouring of popular literature depicted war as both normal and romantic. Wars could be enjoyed vicariously because they would be short and decisive.

In 1906 one of the best-known French professional military pundits, General Bonnal, estimated that the next war would be decided in less than a month; and six years later Commandant Henri Mordacq warned the politicians not to assume that the next war would end after the first great battle—it might last as long as one year.[18] Not only would future wars be short, their conduct would be chivalrous, romantic, and glorious. Anti-war writers of various kinds, including Norman Angell, Bertha von Suttner, and Ivan Bloch faced a difficult task in depicting war's sordid realities to a sedentary, urbanized readership who found 'the spectacle of war even more attractive than the spectacle of football'.

Several years before H. G. Wells coined the famous phrase, writers of future-war fiction had established the myth that the next great conflict would be 'the war to end all wars'. The endlessly repeated term 'decisive' referred not only to victory on the battlefield but included the utopian settlement of international problems. Thus in *Der Weltkrieg* (1904), August Niemann concluded that 'the final aim and object' of Britain's defeat at the hands of Germany, France, and Russia in this 'gigantic universal war' was 'a new division of the possessions of the earth'.

To what extent this extremely popular genre merely reflected existing prejudices or actually moulded public opinion could be endlessly debated. Clarke argues persuasively that the more famous jingoistic authors such as Danrit, Le Queux, and Niemann shared responsibility for the catastrophe of 1914 because they helped to raise the temperature of international disputes. Many others played their part in fomenting and sustaining the self-deception, misunderstanding, and ill will that often clouded relations between the peoples of Europe. By 1914 it is clear that 'the device of the imaginary war had begun to affect international relations in a way never known before'.

In all the multifarious attempts to predict the character of the next great war before 1914 there was almost universal consensus on the belief that it would be short and, necessarily, 'decisive'. As noted above in the case of Mordacq, or Kitchener who in 1914 advised preparing for a three-year conflict, even critics who queried the short-war assumption still underestimated the actual length of the First World War. The elder Moltke's grim speculation about a war lasting seven or thirty years was conveniently forgotten. These predictions were not founded on objective calculations but rather on extrapolations based on the increasing lethality of weapons, by references to the duration of (some) recent conflicts, and, above all, on the understandable fear that the European economy and the political order supported by it could not survive a long attritional struggle. As Schlieffen was to write in retirement after 1906: 'a quick decision is necessary to start the wheels of industry turning again. A strategy of attrition is impossible when the maintenance of armies of millions requires the expenditure of billions.'[19]

Serious students of war, however, were well aware that a long attritional war was very likely indeed in prevailing conditions. On the military side, armies numbered in millions could now be mobilized and maintained by modern transport systems for as long as supplies were obtainable and the 'home front' held firm. In political terms, the rival alliance groups suggested that the war theatres would be numerous and spread over vast areas: provided the alliances were maintained even the complete defeat of one partner would not necessarily end the war. Above all, the propaganda and the rhetoric only slightly distorted the grim reality that the continental powers (less so Britain) were contemplating national wars for the highest political stakes in a pseudo-Darwinian atmosphere of ruthless competition for survival. Though not impossible, the odds after 1900 were heavily stacked against a limited Cabinet War like those which Bismarck had, with difficulty, managed to control.

One writer and anti-war polemicist, above all others, *was* prepared to 'think the unthinkable' by predicting—with a unique combination of statistical data and political imagination—the likely prolonged course of a general European conflict and its catastrophic consequences. His warnings were intensely debated for a few years but had little, if any, tangible influence. Ivan Bloch was a Polish Jew who served the Russian state and made his fortune as a railway contractor, banker, and financier. He published weighty studies on the Russian railways and finance before taking eight years of intensive research to produce his six-volume *magnum opus* on *The War of the Future* in the late 1890s. Only the final volume was published in English in 1899, under the somewhat misleading title *Is War Impossible?*[20] In that year, just before the start of the Boer War, he visited England and over the next two years published articles and lectured to hostile military audiences. He died in January 1902 at the age of 66, so did not live to witness the trial of some of his theories in the Russo-Japanese war of 1904–5, much less the conflict of 1914–18 which he is widely held to have 'predicted'.

It was Bloch's strength, but also his fatal limitation in the eyes of the military and political authorities, that he was a moral crusader (in Tim Travers's apt analogy, 'a Polish Richard Cobden') desiring peace, free trade, and the triumph of commerce and industry over war and militarism. Bloch's study was a staggering achievement for a complete amateur and outsider in military matters. Pre-1914 professional students of war were, with a few exceptions, not the blockheads or arch-reactionaries of popular myth, but they clearly approached the subject as specialists with *parti-pris* assumptions that greatly blinkered their attempts to see ahead. By contrast Bloch brought to the subject a fresh analytical mind steeped in economic, financial, and business matters. Perhaps with some exaggeration, a contemporary scholar has bestowed on his book the accolade of 'the first work of modern operational analysis'.[21]

Bloch's starting-point was that the revolutionary development of fire-power since 1870 had enhanced the superiority of the defender at the tactical level. Improvements such as the magazine rifle, smokeless powder, and flat bullet trajectory would create a lethal fire-swept zone. Both sides would dig extensive trench lines to survive, and the fire zone could only be crossed with an eight-to-one superiority, and then with heavy casualties. The marked superiority of the defensive would result in a long, indecisive, static war: siege conditions would characterize the whole front. There would be vast fortified areas on the model of Plevna, which had withstood a Russian siege for several months in 1877. Moreover, since the embattled armies would be numbered in millions, there would be acute supply problems and little scope for manœuvre. Bloch painted a bleak picture of indecisive attrition:

At first there will be increased slaughter on so terrible a scale as to render it impossible to get troops to push the battle to a decisive issue. . . . Then, instead of a war fought on to the bitter end in a series of decisive battles, we shall have as a substitute a long period of continually increasing strain upon the resources of the combatants.

Everybody will be entrenched in the next war. It will be a great war of entrenchments. The spade will be as indispensable to the soldier as his rifle. . . . Battles will last for days, and in the end it is very doubtful whether any decisive victory can be gained.[22]

He then turned to the social and psychological dimensions of modern war. He perceived, correctly, that a prolonged static war involving the great powers would put enormous and eventually unbearable strain on the economy. He doubted both the soldiers' and the civilians' ability to withstand the stresses of modern war. In addition to psychological stress, mass warfare would cause shortage of labour to produce food, and the railway system would not be able to meet the needs of both the home and military fronts. Anarchists and socialists would foment convulsions in the social order and famine, rather than decisive battlefield victories, would be the ultimate arbiter.

Bloch was too sensible to believe that governments would immediately heed his dire warnings and refrain from embarking on another great war. But he did hope that nations would soon realize the folly of appealing to the arbitrament of war: first, because no definite decision could be speedily secured; and, secondly, because the costs would be ruinous to all parties. War was not literally impossible but resort to it would be suicidal for political, economic, and social reasons.

Like most impassioned propagandists, Bloch tended to overstate what was in some respects a very strong case.[23] The fire zone did indeed pose terrible problems for the attacker but it was not impassable given flexible tactics, close artillery support, and determination. Moreover, it suited Bloch's anti-war stance to stress the fact that technology currently favoured the defender, but he was aware that in time innovation might restore mobility to the battlefield. He was also wrong (or overly pessimistic) on certain significant details. He argued for example that most of the officers would be quickly killed, thus turning the armies into undisciplined mobs. He underrated the medical facilities available for the sick and wounded, and the stoicism and resilience of urban citizens in uniform. In general he failed to make due allowance for such factors as comradeship, patriotism, and military discipline which enabled the Russian armies to keep fighting well into 1917, and the multinational Habsburg armies till the very last weeks of the war in 1918.

Perhaps more surprisingly, Bloch's economic and political views were more orthodox and so far less perceptive than those on military operations where he had no expertise whatever. Like most liberals and pacifists he anticipated that a general war would cause rapid dislocation in trade, commerce, and international credit, but he had no inkling of the ruthlessness with which the belligerent states would interfere in the private sector to take control of manpower, shipping, railways, coal-mines, and everything essential to waging a long, attritional war. Ironically he believed that Russia's self-sufficiency in food supplies and comparative backwardness in industrialization would afford her greater staying power in a long war.

Lastly, and paradoxically, Bloch's pacifist idealism would have stood a better chance of acceptance had the mass of soldiers and civilians of the belligerent states been as physically and mentally unsuited to bear the hardships, deprivation, and suffering of attritional warfare as he assumed. The implications of his treatise were too radical and too disturbing to find military or political endorsement before 1914. His ideas received particularly rough handling from critics in Germany, above all from the distinguished historian Hans Delbrück. He was charged with underplaying the human and moral factors in war, and of completely misrepresenting the place of war in international relations. He was illogical in wanting both disarmament and peace. In sum, Bloch's pacifism was profoundly unpopular in Germany and only a little less so elsewhere.[24]

Graf Schlieffen, chief of the German general staff between 1891 and 1905, studied the same military developments as Bloch, but as a specialist staff officer regarded his task as to prepare for victory rather than to admit that it was unattainable. Whereas Bloch's starting-point was the tactical impasse caused by the enormous development of fire-power, Schlieffen's was the strategic opportunity presented by railroads. Faced by the dilemma of whether to strike first east or west in a two-front war, Schlieffen found a solution in the battle of Cannae in 216 BC in which a Carthaginian army, outnumbered almost two to one, virtually annihilated a larger Roman force by moving round the flanks and into their opponent's rear. In what was termed a double envelopment, the Romans were routed and massacred.[25]

But was such a daring manœuvre repeatable in late nineteenth-century conditions in view of the immensely complicated (and in some respects incalculable) factors involved? These included the rival mobilization time-tables, railroad and other (mostly horsed) transport capacities, enemy reactions, and—not least problematical—the ability of the high command to control and co-ordinate the movements of armies numbering more than a million troops.

The elder Moltke had become increasingly pessimistic about solving this problem by a quick knock-out offensive on either front, but Schlieffen gradually evolved a plan of extreme boldness tinged, one suspects, with desperation.[26] The essentials of Schlieffen's final version of the plan before retirement (dated 28 December 1905) were to concentrate seven-eighths of available forces for a western offensive, the heaviest armies to be positioned in the north between Luxembourg and Aachen with the role of driving through Belgium and the Netherlands, by-passing or enveloping the French northern flank by moving close to the Channel coast, crossing the Seine, and by a giant wheeling movement swinging round to the south-west of Paris. Meanwhile the weaker German armies to the south would wait on the line of the river Meuse, hoping to act as the anvil against which the French forces retreating eastward would be hammered.

Schlieffen allowed approximately six weeks from mobilization for the completion of this vast and hazardous enterprise, on the calculation that

Russia's slow mobilization and poor railways would permit the much weaker German forces in East Prussia to hold out until reinforced by several victorious corps from the West. An essential requirement of the plan was that the French army (and any allies she might have such as the Belgians and the small British Expeditionary Force) must be not merely beaten in battle or forced to retreat but completely destroyed as an organized force capable of resistance.[27]

This is not the place for a prolonged discussion of the issues which generations of historians have debated since 1918, focusing on the question 'was the Schlieffen Plan a blueprint for disaster, or a feasible strategy ruined by the younger Moltke's alterations before the war and his failure as the executive commander in 1914?' Ritter, Bucholz, and others have exposed the weaknesses of the plan itself, as regards for example the shortages of German manpower and heavy dependence on the enemies (Belgium, France, Britain, and Russia) all reacting as anticipated; while Martin van Creveld has argued forcefully that the logistical arrangements foredoomed the plan to failure.[28] On the other hand, a clear-cut negative verdict must be tempered by the facts that, with all its shortcomings, the plan as imple-mented did come very close to success in the West, while outstanding victories were won against Russia in the East.

The most serious charge against Schlieffen and his successor is that they formulated a purely military plan which ignored the political dimension. To achieve 'total victory' in military terms would be difficult enough but, as previous case-studies of the Napoleonic and Bismarck eras have suggested, without a close and continuous political direction it would be virtually impossible. The German general staff's rigorous exclusion of political con-siderations is scarcely comprehensible to the late twentieth-century student. Because it was necessary to his railway timetable Schlieffen pro-posed to invade neutral Belgium and Holland. Violation of the former in particular was almost certain to provoke intervention by the British, as signatories of the Treaty of London (1839), but Schlieffen had no respect for the British Expeditionary Force (which, he remarked sarcastically, could be 'securely billeted' in Antwerp), and he assumed that a quick and total victory would render the Royal Navy's role irrelevant.[29] Between 1905, when Russia was a negligible factor after crushing defeats by Japan on land and at sea accompanied by revolution at home, and 1914 there were many significant changes in the relationships between the likely belligerents in a

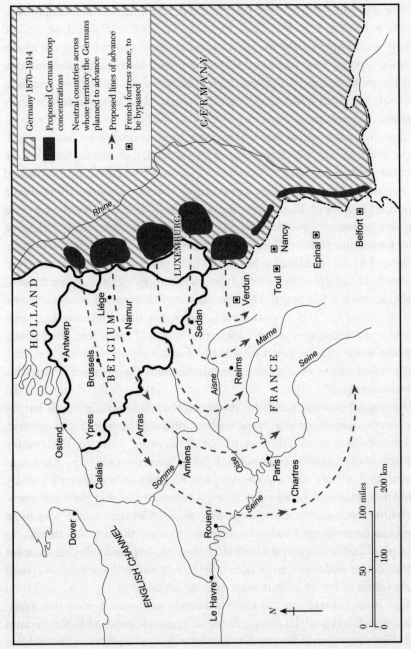

Map 7. The Schlieffen Plan, 1905–1914

general European war and the potential contributions of individual powers, yet the Schlieffen Plan remained unchanged in its essential features. What is truly astonishing is that the plan lacked any political flexibility (let alone alternatives), so that Germany was committed to an immediate all-out attack on France (and Belgium, though Moltke did refrain from invading Holland) no matter where or how the war began. This exemplified a cynical, dynastic eighteenth-century conception of Cabinet War in an era of rampant nationalism, parliamentary institutions, and mass armies.

But if the general staff was guilty of professional narrowness, were not its political masters (or, at least coequals) even more to blame for not laying down clear political guidelines and ensuring that war plans were consonant with foreign policy? There does indeed appear to have been a wide gap between statesmen and military planners. For example, when (in 1900) Privy Councillor Holstein heard that his friend Schlieffen did not propose to be bound by international agreements he replied ('after a long brooding silence'): 'If the chief of staff, if a strategic authority of Schlieffen's stature, considers such a position to be necessary, it is the duty of German diplomacy to adjust to it and prepare for it to the degree that this is possible.' And in trying to exculpate himself after the war had been lost, Bethmann Hollweg wrote: 'During my whole term of office there was never any kind of council of war in which politicians intervened in the pros and cons of the military debate.'[30]

We must, however, avoid the inference that German statesmen before 1914 were unaware of the broad outlines of the Schlieffen Plan, and that had they been informed they would have intervened to secure moderation and the observance of international law. Quite the contrary, 'for some fifteen years, all of Germany's leading statesmen knew and approved of the Schlieffen Plan and of its political implications . . . If the Plan was over-stretching the limits of German power, so was German policy.'[31] In their emphasis on resolution and determination to carry through this incredibly risky plan, politicians and soldiers alike were underplaying the significance of Bismarck's diplomacy in isolating the enemy and preventing third-party intervention in the successful wars of unification.

This leads naturally to the most fundamental flaw of German strategic thinking before 1914: the assumption that if, against great odds, the French and allied forces could be speedily 'annihilated' in an updated version of the battle of Cannae, then battlefield victory would be crowned by a diplomatic settlement so secure as to permit the immediate transfer of the bulk of the German forces to the East. This was extremely doubtful, given that the fundamental strategic aim of the Triple Entente of France, Russia, and Britain, consolidated since 1907, was to force Germany and Austria-Hungary to fight simultaneously on three or more fronts in Europe and to face long-term blockade from the most powerful navy in the world.[32] Without employing the wisdom of hindsight it seems reasonable to suggest that, at the very least, German statesmen and war planners had to ensure that Britain would not intervene until the Schlieffen Plan had had a chance to succeed against France. But since violation of Belgian neutrality was essential to the plan, British intervention had to be considered likely.

In sum, the Schlieffen Plan was implemented under conditions which made its success improbable: even if the French armies suffered severe defeats in the opening battles they had every incentive to fight on in the expectation that their allies would soon put the Central Powers under tremendous pressure on land and at sea. Thus realistic expectations of effective support from allies provides one significant contrast to France's predicament in 1870 and 1940—in the latter case of course Britain was still formally an ally when the French capitulated but was unable to bring succour in land operations. The second contrast with the catastrophes of 1870 and 1940 was that between about 1911 and 1914 France had developed a strong sense of national unity, demonstrated on the outbreak of war in the collapse of socialist plans to disrupt mobilization and the formation of the Union Sacrée.[33] Thus we may conclude that, provided Russia continued to wage war vigorously in the East and the Royal Navy remained dominant at sea, Germany would simply lack the necessary forces to compel France to submit to a draconian peace settlement. Victory, in the full terms conceived by Schlieffen, was not a practical possibility in 1914.

It is a commonplace criticism of First World War commanders that they began with suicidal frontal attacks, were surprised by the onset of static, trench warfare, and were slow in devising methods to break the stalemate, renew mobility, and win the war. While there was some substance in these charges, it did not result from ignoring recent military conflicts. Quite the contrary, for recent wars such as those in South Africa (1899–1902) and the Far East (1904–5) were observed, written up, and discussed assiduously, indeed almost obsessively. Very few professional officers doubted that they were living through a tactical revolution and must try to profit from others' experience. In the critical matter of fire-power, for example, the 1880s witnessed the introduction of magazine-fed rifles, true machine guns like the Maxim, and smokeless powder; while in the 1890s the development of recoil breakers made possible quick-firing guns. Furthermore, although the Russo-Japanese war was limited for both parties, together they had assembled more than two million troops in Manchuria by 1905.[34]

The Boer War was characterized by prolonged sieges, encounter battles, and extremely fluid guerrilla warfare over the vast open spaces of the veld. The Russo-Japanese war witnessed impressive combined operations, the great siege of Port Arthur, wide-ranging manœuvres on a greater scale than Napoleon's, and culminated in a protracted battle of fixed positions around Mukden.

Although most observers, official historians, and other commentators attempted to draw 'lessons' relevant to the next great European war, it should not surprise us that the results were ambivalent or contradictory. A great deal depended on the observer's rank, his specialist interest, what action he was able to witness at first hand, and, above all, which side he was attached to. Among British observers, General Sir Ian Hamilton was unusual in his superior rank and wide previous experience of combat. His two volumes of recollection, *A Staff Officer's Scrapbook*, were favourable to Japan, Britain's ally. But the greatest temptation, to those who found military practice confounded their assumptions, was to say that conditions were so untypical and abnormal that European armies could not benefit from studying them. Thus, for example, in 1911 a German military handbook specifically stated that long-drawn-out frontal battles and siege warfare such as had occurred in the Far East, were most unlikely to occur in a European war.[35] Unfavourable terrain and vegetation were even cited in mitigation of the disappointing performance of both the Russian and Japanese cavalry in the *arme blanche* role, the charge with lance or sabre.

A critical conclusion would be that although there were perceptive individuals who derived accurate insights about the nature of future warfare from pre-1914 conflicts, they made little or no impression on official doctrine which assumed that the next great war *must* be short and that it would be decided by strategic and tactical offensives carried out by soldiers whose high morale and unquenchable spirit would enable them to surmount the hail of lead in the fire zone. But the honest student must admit that all the 'lessons' were open to different interpretations, and that many points which appeared obvious after 1918 were not so beforehand. Indeed, had events turned out differently in 1914, the pre-1914 evidence would have been given quite a different perspective.[36]

In order to appreciate the difficulties which even the best pre-1914 military minds encountered in trying to reconcile the evidence of contemporary warfare with the historical legacy of the Napoleonic era, we may take the case of Commandant Jean Colin, an authority on the development of Napoleon's military genius, whose study *The Transformations of War* was published in London in 1912. One of Colin's most impressive achievements was to demonstrate how the sensible tactical lessons which had been learnt—and promulgated as doctrine—since the Franco-Prussian war were, by about 1900, again in danger of being distorted by peacetime bravado and lack of realism in doctrine and training.[37] Hence the most up-to-date lessons from Manchuria and Africa would be a salutary corrective—if heeded.

Contrary to the tone of much popular discussion, Colin pointed out that the Boers had eventually been beaten by the British despite their initial advantages in numbers, familiarity with the terrain, and surprise. But as an irregular force the Boers lacked discipline, cohesion, and the offensive spirit. On both the tactical and strategic levels they had failed to press home their early successes. British tactics had initially been suicidal and they had adapted only slowly in comparison with the Japanese, who, for example, kept up their covering artillery fire until the infantry entered the enemy trenches.[38]

Colin predicted that future battles would result from both armies acting offensively. He noted the size of armies had recently increased enormously and also that there was a tendency for battles to last ever longer—from an average of about five hours in the mid-eighteenth century, to two days in some of Napoleon's greatest battles (Wagram, Bautzen, Leipzig), and even longer in Manchuria where the Japanese turning movements at Mukden had taken six days. How could armies which would be so immense as to occupy the whole theatre of operations nevertheless be able to manœuvre and force a decision? Colin answered that the problem of huge numbers would be offset by improved roads and railways while the commander would have the advantage of telephones and wireless telegraphy. Moreover, balloons and aeroplanes would soon replace cavalry in the reconnaissance role. He argued that, although industrial and military progress favoured the defensive in frontal fights, the offensive could impose battle and gain victory: 'for the assailing army occupies the whole theatre of operations and sweeps all away on its passage.'[39]

Colin perceived that in the Russo-Japanese war both sides were nearing exhaustion when peace terms were mediated by the United States, and that they had allowed this to happen because the interests at stake were insufficient to prolong the struggle. But, he added, such limiting considerations would never apply in Europe:

Without speaking of the passions that would animate most of the belligerents, the material conditions of modern war no longer admit of avoidance of the radical decision by battle. The two armies occupying the whole area of the theatre of operations march towards each other, and there is no issue but victory. It is impossible to avoid the encounter, impossible also to seek in it but a half-success.[40]

Yet beneath all the sensible comments on the implications of the wars in South Africa and Manchuria, Colin's underlying theme was the continued relevance of the principles and spirit of Napoleonic warfare, to which he devoted approximately one hundred pages in the latter part of his study. New weapons, larger masses of troops, and more efficient means of transport must necessarily cause modifications in applying Napoleonic methods; some principles might even have fallen into disuse: 'Nevertheless, for him who knows better than to copy forms slavishly, it will still be in Napoleonic war that he will find the models that should inspire, the subjects that should be meditated, and the ideas that should be applied in the twentieth century.'[41]

Until quite recently it was difficult to find a historian or political scientist with a favourable word to say in explanation or mitigation of the pre-1914 'cult of the offensive', since the doctrine seemed directly responsible for the mass slaughter in the opening months of the war. Foch, Grandmaison, and other proponents of French military doctrine certainly had a lot to answer for: they assumed too easily that the Napoleonic model of the decisive attack could be repeated in modern conditions; they underplayed the significance of trenches and fortifications; implied that those who seek battle always win; and that a quick military victory will inexorably lead to a triumphant political settlement.[42] Common sense suggests, however, that the terms 'offensive' and 'defensive' have been falsely viewed, both by the French theorists and by some later commentators, as irreconcilable alternatives. Forces launched upon a strategic offensive will inevitably have to apply defensive tactics, while their opponents, initially forced on to the defensive, will sooner or later have to counter-attack.

Douglas Porch began a partial rehabilitation of French military offensive doctrine by placing it firmly in the context of domestic politics in the bitter aftermath of the Dreyfus affair. The doctrine became more intelligible when linked to the army's protective measures against radical attacks; the issues of the length of conscript service and the reliablility of the reserves; and the shortage of up-to-date artillery. On the strategic plane, France would have to take the offensive, though she was most unlikely to initiate hostilities, to fulfil her obligations to Russia.[43]

Michael Howard and Azar Gat have developed the argument further.[44] Foch, Colin, and other military theorists recognized that extended en-trenched positions were virtually impregnable and that the fire zone was wider and more lethal than in the past, but they rejected the inference that aggressive, offensive strategy was no longer feasible. They feared that excessive emphasis on fire-power and the advantages of the defensive was a recipe for another disaster like that of 1870. They therefore put more stress on war's strategic and moral aspects, and found justification for confidence in emulating the Japanese performance against the Russians, in which the former's offensive spirit had been crucial.

Unlike the South African war, which might be dismissed as a testing ground because the adversaries were so ill-matched, the Russo-Japanese war involved two major powers equipped with the most modern weapons: magazine rifles, machine guns, and quick-firing field artillery. The Russian trench lines at Port Arthur and Mukden were protected by barbed wire and electrically detonated minefields. Both armies were equipped with the telegraph and field telephones. The Japanese army had been trained by the Germans and her navy by the British. This was precisely the kind of conflict which Bloch had foreseen to be impossible—or at least suicidal. The costs were indeed extremely high; for example, the Japanese lost some 50,000 troops in their assaults on Port Arthur and 73,000 in the ten-day battle at Mukden. But the main lesson seemed to be that, despite the formidable advantages of the defensive, it was the attacker, vigorous and prepared to make sacrifices, who prevailed. Superior morale seemed to be at least as important as weapons. As one French commentator wrote in 1906: 'The victory will go to the offensive which stimulates moral forces, disconcerts the enemy, and deprives him of his freedom of action.'[45] The Japanese

forces had set standards in sophisticated tactics, determination, and willingness to accept casualties which any great power would have to emulate in
event of a general European war. Victory had confirmed Japan's status as a
great power, whereas Russia had suffered defeat and a revolution.

In August 1914 the French armies showed that they possessed the qualities of offensive *élan* and self-sacrifice in abundance. In six weeks they
suffered some 385,000 casualties, of whom 110,000 were killed. But, as
Michael Howard has suggested,[46] it was not necessarily that Plan XVII (the
final French offensive plan adopted in 1911) or the accompanying offensive
spirit were entirely to blame; unfortunately, the French army was also
tactically inept. It lacked heavy artillery; close co-operation between infantry and artillery was poor, and fieldcraft—not helped by colourful uniforms—deplorable. Most of the French losses occurred not in set-piece
attacks against barbed-wire entanglements and trenches, but in encounter
battles when both armies were on the move and the French were destroyed
by artillery and machine-gun fire. Thus we may conclude that the offensive
doctrine was not completely responsible for the terrible losses of 1914 as has
often been assumed. The most insidious error in pre-war emphasis on the
offensive was that it would produce 'decisive' success on the battlefield
which would promptly end the war in a 'decisive' victory.

From its inception, the First World War was 'about' significant political
issues and a test of power relationships between the belligerents, but it was
only Clausewitzian in a very loose sense. With the possible exception of
Austria-Hungary, whose war aims against Serbia were made brutally clear in
her ultimatum, the other participants were prompted by complex considerations of territorial ambitions, alliance obligations, and fear of the
penalties of non-participation. Once hostilities began, war aims sprang up,
mushroom fashion, even in such minor belligerent kingdoms as Bavaria and
Saxony.[47] Though 'militarism', as defined by Gerhard Ritter, varied considerably between Germany at one extreme and Britain at the other, it was
manifest in every belligerent power, in the sense that nowhere were the
armed forces under the complete control of civilian governments, and
nowhere were there satisfactory institutional arrangements to preserve
civil—military harmony in the evolution and control of strategy.[48] This
governmental failure had serious consequences everywhere, but was disastrous in the cases of Germany and Austria-Hungary.

Thus military leaders everywhere enjoyed a remarkable amount of autonomy, such as Moltke had vainly sought in 1870. But there were certain unwritten obligations which might be summarized as follows:[49] the statesmen expected the generals to produce a speedy victory; the war must, if possible be fought on enemy territory; national resources, especially in manpower, would be made available for the short period anticipated (months rather than years); and the armed services had to achieve their objectives or forfeit a great deal of their power and prestige. In other words, the military leaders had to be able to deliver victory in a short time and at acceptable cost to justify their expense to the nation. Consequently, while it was still possible in Britain—a *status quo* power mainly dependent on her naval strength and economic pressure in a long war—for Kitchener to warn at the outset that the government must prepare for a three-year struggle, the continental warlords had no option but to plan short, offensive campaigns culminating in decisive battles and victory. Bloch's grim depiction of trench stalemate, attrition of manpower and resources, accompanied by mounting pressure on the home fronts, was simply not acceptable.

Nor of course was there much realistic reflection about what lay ahead on the part of the hundreds of thousands who greeted the declaration of war with a mixture of joy, relief, idealism, escapism, and self-delusion. In Germany, for example, the declaration of war released a heady excitement that swept the whole country. To the young 'it offered an exhilarating holiday from the dull routines of normal life . . . and a promise of self-fulfilment'. There was a widespread belief that war would end party strife and restore humane, civilized values: it would cleanse mankind from all its impurities.[50] These idealistic expectations had been fostered by competitive nationalism over the past generation without the corrective of realistic depiction of war in art, literature, or history. Caroline Playne captured something of this euphoric spirit among the crowds in August 1914 when she wrote: 'At last! At last! We get what we want, we can do and die! And they felt in their hearts, intense relief that there was to be no more negotiating, no more thinking, no more heeding, only rushing on, on, gloriously, splendidly on, all traces kicked over, all bridles thrown away!'[51]

Doubtless only a small segment of the total population in any of the belligerent nations reacted in this hysterical way and of these many would be civilians whose own lives were not at risk. But those who openly opposed the war were probably an even smaller minority. Nationalist propaganda and literary euphemisms had obscured for most people the horrific realities of warfare.[52] So, no matter how limited the enthusiasm for personal sacrifice might be, most citizens were prepared to do their duty. In Michael Howard's words 'They trusted their rulers and marched when they were told.'[53] Their remarkable stoical endurance after the early romantic illusions had been shattered goes far to explain why Bloch and other prophets of rapid disintegration were proved so very mistaken.

So powerful in the West are the literary, pictorial, photographic, and film images of static attritional warfare between unbroken trench lines in a devastated landscape that it requires an imaginative effort to recall that operations began with the great war of movement and encounter battles which Colin and others had predicted. In the West the trench stalemate did not occur for several months, and in the East not at all, in that the battle lines continued to be fluid. The Germans achieved spectacular victories in East Prussia and, in the West and with stronger and more flexible leadership, might have extended their advance even further than the Marne.[54] Neither side was able to avoid the delusion that 1915 would bring the 'decisive victory' which had not been achieved by the first Christmas. As no single belligerent, not even tiny Belgium, had been completely overrun and defeated, the alliance system provided insurance and encouragement to hope for eventual victory. Thus 'the short war became impossible not at the Marne but at Tannenberg and Mons, when the Russians and British demonstrated that they intended to fulfill their alliance commitments to the French'.[55]

Moreover, since governments everywhere had surrendered power to the generals (and admirals) to win the war, they were now obliged to give the latter another chance to deliver or prove themselves incompetent. Furthermore, since pre-war provision of ammunition and other vital military supplies had been based on a short war of movement, there was now bound to be a long interval—of at least a year—to allow for mobilization in depth, not to mention the unscrambling of the profound muddles of 1914 based on short-war assumptions. But the very decision to mobilize manpower and resources on a massive scale that amounted to 'total war' made a negotiated compromise peace without an outright victor all the more unlikely. Finally, we should not assume that the military leaders everywhere came to accept that total victory, as conceived before 1914, had been proved to be unattainable due to the superiority of the means available to the defence. Quite the contrary, German leaders (and others) continued to struggle throughout the war to achieve the decisive victory which would result from superior material strength and moral qualities.[56] Victory, on the lines of Moltke the elder's achievements, and given new credibility by the Japanese successes against Russia on land and sea, proved elusive in 1914 but it was not yet demonstrably 'an illusion'.

The Pursuit of Victory in the First World War and the Aftermath

It has long been unfashionable to write about the First World War in terms of victory and defeat. Military historians know there was a decisive outcome on the Western Front in 1918, and most have given credit to the Allied armies, and notably the British and Dominion forces, for the remarkably successful offensive operations which resulted in the armistice on 11 November.[1] But for general students of the war winning and losing have been overshadowed by other considerations. Would the governments which so readily went to war in 1914 have done so had they been able to foresee its duration and terrible costs? Why was the conflict not ended sooner by a negotiated 'peace without victory'? How did the comparatively modest, or at least specific, war aims of 1914 become the unlimited and vague ambitions of 1918? Above all, why did the peace settlements prove to be so brittle and ephemeral, making a mockery of the rhetoric of a 'world safe for democracy' and of 'a war to end all wars'? It has consequently been difficult, even for military historians, to regard this conflict in Clausewitzian terms, that is as an instrument of policy used to secure specific objectives under political control and direction. On the contrary, its clumsy direction, appalling nature, confused ending, and short-lived results have made it a byword for incompetence and futility. While it will probably be impossible to overthrow these widely held opinions, based as they are on hindsight, moral revulsion, and an assumption that things could have been done differently, the view advanced here is that the conflict *was* about important issues, that it *was* eventually ended by military means, and that, despite all the disappointments, victory was much preferable to defeat.

The failure of the Schlieffen Plan to deal a knock-out blow on either front in 1914 and the acceptance by both sides of an attritional siege war on a vast scale, in the winter of 1914–15, ensured that it would be a long struggle, since military and civil mobilization for a 'total' effort would take at least a year. In principle, and in terms of Clausewitzian theory, it should have been possible to negotiate a peace settlement before the war of attrition developed, say by the end of 1915. Yet the vast historiography on wartime diplomacy does not suggest that a clear-cut opportunity was missed. Russia might well have been willing to negotiate had the Germans been interested in moderate terms. But the latter insisted on a 'final victory which would destroy all the forces opposing them'. They looked to 'a victory which would give them a final peace'.[2]

In any case governments everywhere had yielded far too much authority to the generals, who were confident (or appeared to be) that one more offensive would bring victory. Given the prevailing superiority of defensive weapons and tactics, the military instrument (both on land and at sea) proved to be inflexible and indecisive. The very number of 'players' in the conflict and the addition of new ones to the 'game' (such as Italy and Bulgaria in 1915, Romania in 1916) also militated against a negotiated peace without victory.[3] Not least significant, the peoples of the warring nations proved to be far more resilient, determined, and bellicose than the prophets of a rapid collapse of morale, such as Bloch, had expected. Even the terrible bloodletting of 1914 and 1915 did not cause disillusion. This, in Michael Howard's words, 'was exactly that trial of patriotism, manliness, and endurance for which the nations of Europe had been preparing themselves for a generation'.[4] Even when deadlock had clearly been reached in 1916 only a tiny and generally unpopular minority advocated a compromise peace.

Even the staunchest defenders of the military leaders of the First World War would accept that strategy and tactics, particularly in the first half of the war, were often crude and costly. The debate now is rather concerned with how quickly they were willing and able to adapt, and how much credit they deserve for eventually breaking the trench deadlock and restoring comparatively mobile operations.[5] What should not surprise us, however, is the determination (or obstinacy) with which commanders pursued the will-o'-the-wisp of victory. Numerous generals deemed to be lacking the will to win, or simply unable to deliver the expected results, such as Moltke, Falkenhayn, Joffre, Nivelle, French, and Ian Hamilton, were dismissed, or 'degommé' in the idiom of the time. The survivors, such as Foch, Mangin, Haig, or Allenby, were noted for their determination to secure victory without agonizing over the human cost. Even the more defence-oriented Pétain was committed to eventual victory, though prepared to wait for a vast influx of Americans to deliver it.

As national war efforts became more 'total' and war aims broadened correspondingly, so politicians looked for more vigorous war leaders, like Lloyd George and Clemenceau, rather than moderates who would be willing to accept—or at least consider—more moderate peace terms. Germany got the best (and worst) of both worlds from August 1916 onwards in the 'silent' or thinly veiled dictatorship of Generals Paul von Hindenburg and Erich von Ludendorff, whose effectual control of the civil government was tacitly recognized by the Reichstag in October.[6] They broadly endorsed, and even expanded, the ambitious territorial claims in both Eastern and Western Europe as adumbrated in the 'September programme' of 1914. So long as they remained in effective control of the government and military victory remained a possibility there was no hope of a negotiated peace.

Far from the initial failure of either side to end the war by Christmas 1914 being conducive to a negotiated settlement, there was a general tendency towards escalation of both the military effort and of war aims. Having made substantial gains in both East and West, the Germans saw no need to compromise, whereas for France and Britain it was inconceivable to abandon the struggle from such a position of inferiority. To such original war aims as the recovery of Alsace-Lorraine and the liberation of Belgium there were added more idealistic goals such as self-determination for the peoples of Central Europe and the crushing of Prussian militarism. 'The flower of British and French manhood had not flocked to the colours in 1914 to die for the balance of power.'[7] But beneath the confused surface of conflicting war aims there remained a constant issue of grand strategy: Germany's drive for hegemony in the centre of Europe. For this reason, even this 'most frustrating and intractable of conflicts conformed with Clausewitz's dictum that "war . . . is a true political instrument, a continuation of political activity by other means" '.[8] However, with the wisdom of hindsight, and from a global perspective, this ferocious European civil war had the ironic effect of ending Europe's dominance by a massive transfer of power and influence to the United States and the Soviet Union.

By 1917 war weariness was manifest everywhere. Though Russia remained at war, the February Revolution revealed widespread unrest in the cities and deteriorating morale in the armed forces. In May mutinies affected a large number of French divisions after the failure of Nivelle's overambitious offensive, and the home front came as close to manifesting defeatism as at any time in the war.[9] In July the Reichstag passed a Peace Resolution in favour of a moderate settlement without annexations or indemnities, but its only practical outcome was the dismissal of Bethmann Hollweg and the confirmation of Ludendorff in supreme authority. A papal peace message in August also foundered on Germany's unwillingness to define her territorial claims, particularly regarding Belgium. Lastly, on 29 November, the Conservative elder statesman, Lord Lansdowne published a letter in the *Daily Telegraph* advocating a negotiated peace to save Europe from anarchy and revolution. None of these initiatives, however, could make any headway against the incompatible minimum war aims on each side.[10] Only outright victory, it seemed, could justify national sacrifices and yield dividends in the form of territorial gains and indemnities. Lenin and his followers did differ fundamentally in working for the total defeat of their own country, but this bizarre strategy was based on the false expectation that a European-wide Communist revolution would ensue as a direct result. By the end of 1917 the United States' involvement and Russia's imminent withdrawal raised hopes for an idealistic victory which would create a new Europe and a greatly improved international system to replace the traditional balance of power adjusted by frequent conflicts.

What would 'victory' mean for the various belligerents? As A. J. P. Taylor has pointed out, most of them began the war with vague or negative aims and tended to rely on victory itself to provide retrospective justification and goals. For the Entente the essential war aim was to survive as independent great powers, but since this lacked popular appeal the struggle had to become one for democracy and the destruction of Germany as a militaristic autocracy. From the outset it became clear to the rulers of the Central Powers and Turkey that they would face domestic upheaval if defeated. The Russian Revolutions in 1917, the harsh terms imposed on her by Germany, and the murder of her imperial family offered a terrible warning. Victory might preserve their social systems and political institutions; defeat would entail revolution and punitive peace terms. Consequently, for the Central Powers and their allies, victory came to be regarded as an end in itself, 'the key to all problems, domestic and external'.[11]

In December 1916 the Central Powers had opened the diplomatic struggle over war aims and domestic opinion with their Peace Note, deliberately issued a few days after victory over Romania. The note, belligerent in tone, did not specify war aims but offered vague conditions to form the basis of a lasting peace. It met with a vehement and equally hostile collective reply from the Entente. In reality Bethmann Hollweg's conditions, though quite severe, were far milder than those entertained by Hindenburg and Ludendorff. The generals opposed any territorial concession to France, required Luxembourg to be annexed, and the Belgian and Polish economies to be subordinated to Germany's. They also intended to annex a strip of territory on Poland's eastern border.[12]

The Entente's uncompromising reply on 10 January 1917 can by viewed primarily as a bid for the ideological sympathy of the United States. Belgium, Serbia, and Montenegro were to be evacuated, as was also all German-occupied territory in France, Russia, and Romania. Most significantly, Britain and France demanded that Europe be reorganized on the basis of nationalities, thus encouraging internal revolution and disintegration in the Habsburg and Ottoman Empires. It was a remarkably bold declaration given the Entente's weak strategic position at the time. Indeed, a peace imposed by the Allied coalition in February 1917, before the United States became a belligerent, 'would have been much harsher than the treaty that eventually emerged at Versailles'.[13]

President Woodrow Wilson's famous 'Fourteen Points', set out in an address to Congress on 8 January 1918, reveal how far he had moved away from the position of a 'peace without victory' enunciated a year earlier. As a response to the Soviet demand for open diplomacy Wilson appealed to idealist, progressive opinion in his own country and in Europe. He echoed Lloyd George's earlier call for national self-determination, justice, open diplomacy, and the creation of a League of Nations. Shorn of their qualifications, both Lloyd George's and Wilson's statement of war aims indicated a determination to destroy the German and Ottoman Empires and to reconstitute the Habsburg Empire on federal principles. The fourteen points, despite their essentially contradictory character, are deservedly famous, because they provided the basis for the armistice in November 1918, and formed the initial terms of reference for the subsequent peace conference, but in the short term they were more successful in influencing Allied public opinion than in altering government policy.[14]

At the beginning of 1918 the strategic situation looked extremely grim for the Entente; indeed, Allenby's capture of Jerusalem just before Christmas was almost the only encouraging event. The French army had apparently recovered from the mutinies of the previous spring, but was of dubious quality as an offensive force. The long, attritional campaign of Third Ypres appeared to have exhausted the British as much as the enemy. This was demonstrated when initial success in a tank-led offensive at Cambrai in November ended, like so many previous operations, with an enemy counter-attack driving the survivors back to their start line. Meanwhile the Italians had suffered a near-fatal defeat at Caporetto; and Romania had been forced to seek an armistice. But by far the most disturbing event was the Bolsheviks' seizure of power and Russia's imminent departure from the war. If, as estimated, Germany could transfer up to 50 divisions to the Western Front, she would stand a good chance of winning before American intervention on land became effective. On the German side, Ludendorff staked everything on this last, desperate offensive, confident that his use of surprise, a brief preparatory bombardment, and new infiltration tactics by specially trained storm troopers would succeed in March 1918 where Moltke had failed in August 1914.[15]

Ludendorff's final series of offensives, beginning on 21 March, did indeed come uncomfortably close to victory, in the sense that a complete breakthrough was achieved on the British Fifth Army's front; the attackers penetrated to the outskirts of Amiens; and thereby nearly caused the hoped-for split between the Allies, with Haig contemplating a retreat to the Channel coast.[16] Instead, at the height of the crisis, Haig and Pétain rallied to the common cause, agreeing at the Doullens conference that henceforth Foch would act as co-ordinator of their armies (and the American forces) on the Western Front. Ludendorff's subsequent attacks in April and May did not achieve the same degree of surprise or penetration, but the crisis was far from over. As late as 27 May the French defences buckled in face of a German onslaught in Champagne, and by the end of the month the attackers had once again reached the river Marne.[17]

This disaster caused British leaders to fear the consequences of a French capitulation followed by that of Italy. At a meeting of the Imperial War Cabinet on 11 June, Lloyd George surveyed the situation in terms of black despair. 'The possibility of complete defeat in France had to be considered . . . with the British Empire and the United States left standing alone . . .'—full attention must therefore be given to intervention in Russia as one means of carrying on the war.[18] As recent research has emphasized,[19]

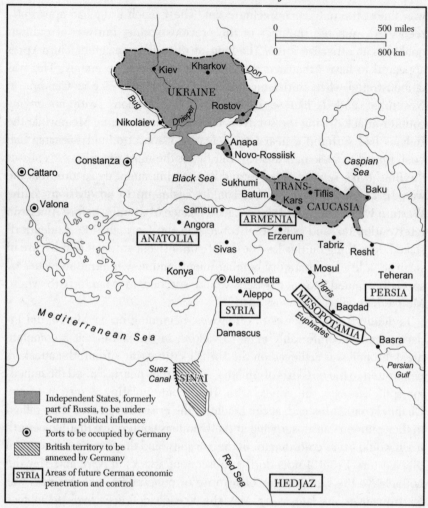

Map 8. Allied fears of German expansion in the East, 1917–1918

throughout the summer of 1918 senior members of the British Government, including Milner and Curzon, as well as the Chief of the Imperial
General Staff, Sir Henry Wilson, were preoccupied with the danger that
Germany would achieve a victory or stalemate on the Western Front and
would drive south-east through Russia to threaten India and British interests throughout the Middle East. Short of that nightmarish prospect, if
Germany could seize and retain control of the Caucasus she would be

immensely strengthened in access to raw materials and would be immune to British seapower. Curzon in particular believed that Germany's main aim was to destroy the British Empire. On 25 June 1918 he told the Imperial War Cabinet that 'she (Germany) can afford to give up everything she has won in western parts, in France and Flanders, if only this door in the east remains open to her'.[20] He and some other members of the Imperial War Cabinet, could not envisage the war in the West terminating before 1919 or 1920, by which time Germany would have consolidated her position in the East at Russia's expense.

On 25 July 1918 Sir Henry Wilson presented a memorandum to the Imperial War Cabinet on 'British Military Policy, 1918–1919', in which the most optimistic of five scenarios was that the current German offensive would be halted; the other four outlined varying degrees of Allied defeat and disunity. In summarizing the ministers' reactions, Wilson noted that only the Australian Prime Minister, Billy Hughes, believed 'the Boches' could and should be beaten in the West. Lloyd George, Borden, Smuts, Massey, and Milner were all sceptical about any kind of victory over Germany 'and so they go wandering about looking for laurels'.[21]

The Allied generals were no more optimistic about an early victory in the West. Three days after the dramatic Allied success in the battle east of Amiens on 8 August (which Ludendorff in his *Memoirs* was to call 'the black day of the German army in the history of this war'), Henry Wilson recorded Foch's plans to seize the German lateral railway between Lille and Metz in 1919. But at dinner on 11 August Haig said 'we ought to hit the Boche now as hard as we could, then try and get peace this autumn'.[22]

Haig's optimism that 'one more push' would bring victory in 1917 had proved mistaken and left him open to criticism, but in the autumn of 1918, though still occasionally hedging his bets, he was much more perceptive than Henry Wilson, Foch, and the politicians. On 10 September, for example, he recorded that in the past month his armies had captured 77,000 prisoners and nearly 800 guns. There had never been, he wrote, such a victory 'in the annals of Britain'. German discipline was fast disappearing. 'It seems to me to be the beginning of the end.' He told General Jack Seely, a former Secretary of State for War (now in the Munitions Department), that they ought to aim at finishing the war then and not to delay the provision of tanks until designs were tested. On 1 October the Third and Fourth Army commanders, Byng and Rawlinson, confirmed Haig's hunch that only con-

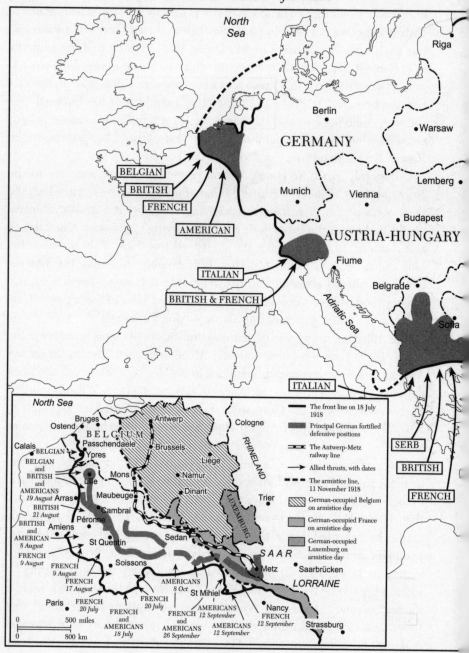

Map 9. Allied victory on the Western Front in 1918

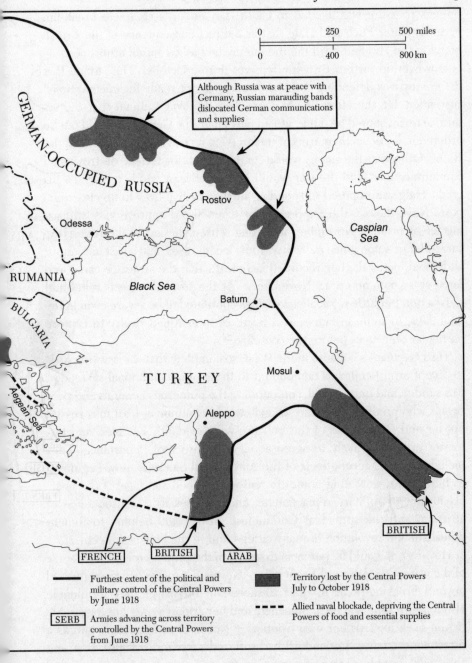

0   250   500 miles

0   400   800 km

Although Russia was at peace with
Germany, Russian marauding bands
dislocated German communications
and supplies

GERMAN-OCCUPIED RUSSIA

Rostov

Odessa

Caspian
Sea

RUMANIA

Black Sea

Batum

BULGARIA

TURKEY

Mosul •

Aleppo

Aegean Sea

BRITISH

FRENCH   BRITISH   ARAB

—— Furthest extent of the political and
military control of the Central Powers
by June 1918

SERB Armies advancing across territory
controlled by the Central Powers
from June 1918

█ Territory lost by the Central Powers
July to October 1918

- - - Allied naval blockade, depriving the Central
Powers of food and essential supplies

tinuous pressure was needed to cause they enemy's complete break-up. Five days later Foch told Haig that as a direct consequence of the British piercing the Hindenburg Line the enemy had asked for an armistice.

However, in further British offensives during October, Haig found that the enemy was defending stubbornly and was not ready for unconditional surrender. He therefore wisely counselled the War Cabinet that the armistice terms should be moderate; in particular, the Germans should not be ordered to lay down their arms or be forced to retreat to the east bank of the Rhine. Otherwise the enemy would conduct a fighting retreat, destroying all communications, and the war would last for at least another year. To his credit Haig was opposed to spending further British lives to no clear purpose. Nevertheless, the Supreme War Council did approve very stiff armistice terms on 1 November, including retirement east of the Rhine, the surrender of 5,000 guns, railway wagons, locomotives, and a vast amount of additional material. Haig recorded laconically that the armistice came into force at 11.00 a.m. on 11 November.[23] At the front there was relief and satisfaction but little rejoicing: casualties continued to be severe even in the final days, in some instances as a result of inter-Allied rivalry to capture particular objectives before the ceasefire.[24]

Thus Germany's collapse in the West, so complete that she was prepared to accept armistice terms tantamount to those of unconditional surrender, was sudden and unexpected, confounding the numerous generals and politicians who predicted that the struggle would continue at least into 1919 if not beyond. The causes of this collapse were evidently complex, and historians are never likely to agree about their order of importance. They include the long-term effects of the Allied naval blockade, now beginning to bite and exacerbating domestic reasons for food shortages;[25] the near-exhaustion of military replacements and supplies as a consequence of attrition; and the spread of Communist propaganda helping to foment defeatism and revolution in many cities—and in the high seas fleet.

However, it would be perverse to question the primary role of the course of the war on the Western Front where Germany's forces had been pushed beyond breaking-point in their unsuccessful offensives; her formidable Hindenburg Line had been breached; and her armies slowly but inexorably driven back towards her own frontiers.[26] Not least as important, and as a direct consequence, Ludendorff had suffered a nervous collapse. Between July and November some 385,000 German prisoners had been taken and over 6,000 guns captured—with the British and Dominion divisions playing the leading role. Also the fact that the total of American troops in France had reached two million when the war ended can hardly have helped German morale.

Some historians, such as Marc Ferro, have suggested that the 'eastern' war theatres had eventually proved worthwhile, in that Turkey's and Bulgaria's surrenders preceded those of the Central Powers and presumably influenced them; but others, such as Trevor Wilson, point out that Germany's impending military collapse had been evident since at least the 'black day' of Amiens on 8 August, and that it was the knowledge of Germany's approaching doom which fatally demoralized her lesser allies.[27]

The chronology of who influenced whom is probably impossible to establish, but what needs to be stressed is that in all the theatres concerned clearcut, and in some cases, dramatic military victories, led to the ending of hostilities. Thus at Amiens on 8 August, Rawlinson's Fourth Army, with Canadian and Australian contingents playing a prominent role, and employing more than 600 tanks and powerful air support, broke through the German front and advanced about 6 miles. The French army also made a significant contribution to the Allied counter-offensive, notably in the action at Montdidier. The enemy was pushed back to his start line of March 1918. On 12 September the first all-American offensive obliterated the notorious St Mihiel salient, and at the end of the month a predominantly British force breached the Hindenburg Line.

For eight months following Allenby's capture of Jerusalem in December 1917 the Palestine front had been 'all quiet'. But on 19 September 1918 Allenby opened what proved to be the decisive offensive with tremendous advantages, including command of the air, overwhelming superiority in cavalry and excellent intelligence.[28] He was faced by German-led Turkish forces whose morale was very low. The mobile operations which began at Megiddo and culminated in the capture of Damascus by Australian cavalry and Arab irregulars on 1 October would become a much-studied model at British military institutions between the wars. Turkey capitulated before the end of October. Meanwhile, Bulgaria had withdrawn from the war on 30 September following a hazardous mountain offensive on the part of the Allied armies in Salonika commanded by Franchet d'Espérey. British, Serb, Greek, and French forces were suffering heavy casualties and appeared to be held up when the Bulgarian command took the fatal decision to make a tactical withdrawal. This resulted in a chaotic retreat and the loss of Skopje after a dramatic 60-mile dash by the French colonial forces. Surrender followed promptly and the remainder of the country was occupied.[29] Italy too had won the decisive victory of Vittorio Veneto over Austria before the ceasefire on the Western Front.

By far the most important defeat—despite the British public's persistent refusal to take pride in it[30]—was that inflicted on the German armies on the Western Front. Although the Allied advance lacked the drama of a decisive breakthrough and pursuit, it nevertheless skilfully outmanœuvered the stubborn defenders from a series of river and canal lines on which Ludendorff had hoped to stabilize the front during the winter. In the far north, for example, the Second and Fifth British Armies occupied the Belgian coast, entered Lille on 17 October, and three days later crossed the river Lys. By the beginning of November they had pushed the defenders across the river Scheldt. Further south, British First, Third, and Fourth Armies forced the crossing of the river Selle, despite appalling weather conditions and very strong defences. A further advance of 6 miles yielded the town of Le Cateau and 20,000 German prisoners. The loss of these strong defensive positions, coinciding with an impressive American advance in the Argonne, caused Ludendorff's resignation on 26 October.[31] In the remaining days of hostilities the Fourth Army crossed the Sambre canal on a front of 15 miles, and the stage was set for a great battle on the Germans' last defended river line of the Meuse. However, the armistice occurred before this could be fought to a finish, and by a neat twist of history, British units entered Mons, where the British Expeditionary Force had first encountered the Germans in 1914, on the final morning of the war.

As will be clear from the preceding summary, the Allied advance during the final three months had been steady but not dramatic compared, say, with the German advance in March. Despite low morale, the German defenders fought stubbornly against superior artillery assisted by dominant air forces. There were a few incidents of large-scale surrenders but never enough to cause disintegration on a wide front. Taking the whole front over the period of three months, the total advance had averaged about 60 miles, a rate of considerably less than one mile per day. That the German army was in the process of being beaten is demonstrated by Ludendorff's resignation, and by the acceptance of armistice terms which were so severe that they precluded any hope of resuming the struggle.[32]

On the other hand, the German army retained its basic order and discipline: there was no disintegration into anarchy such as had affected the heavily politicized Russian forces in 1917. Also of vital importance for the flourishing of myth, Germany had wisely signed the armistice before more than a tiny portion of her territory had been fought over. Had Germany's disintegrating armies been relentlessly pursued across the Rhine it would have enhanced the authority of the victors at the peace conference. Finally, the Allied ceasefire terms, though severe, fell short of the truly punitive. Berlin and other large cities were not occupied and the German soldiers were not detained as prisoners or used as slave labour as the Germans themselves had used *their* prisoners in some occupied countries. Instead they marched home in formed units with bands playing and carrying their personal arms. Numerous eyewitness accounts show that returning units received an emotional welcome more appropriate to victors.[33] Hence the propaganda myth of the political 'stab in the back' of an undefeated army. In reality the war ended chaotically for Germany, with naval and military mutinies, revolution in Bavaria and Berlin, and the Kaiser forced to abdicate to prevent a Communist seizure of power.

An important consequence of the German defeat in the West was that the draconian terms of the treaty imposed on Russia at Brest-Litovsk on 3 March 1918 were largely nullified. By this treaty Russia had yielded sovereignty over territory west of a line from the Gulf of Riga to Brest-Litovsk, including Poland, Courland, and Lithuania. Estonia and Livonia were to remain under occupation until 'proper national institutions had been established'. Russia was required to evacuate the Ukraine and Finland, disarm its navy, and demobilize its army. Russia thereby lost sovereignty over 34 per cent of its pre-1914 population (some 55 million people), 32 per cent of its agricultural land, 73 per cent of its iron-ore output, and 89 per cent of its coal. These brutal terms convey some idea of what defeat could have entailed for the West, and also put a somewhat different complexion on the supposedly unjust and vindictive terms of the Treaty of Versailles.[34]

So much idealistic expectation was placed in the post-1918 peace settlements that they were bound to prove disappointing. In fact it has remained fashionable to heap criticism upon the peacemakers and the terms they drew up, on the assumption that the latter were directly responsible for the numerous conflicts that followed. But, as Kalevi Holsti has reminded us, 'The task of creating a durable peace in the aftermath of the First World

War would have been beyond the intellectual and diplomatic capacities of most mortals.' The power vacuum caused by imperial disintegration in Central Europe, the long civil war in Russia, and the stirrings of nationalism in Ireland, the former Ottoman Empire, and further afield in Africa and Asia, would have led to bloodshed whatever was decided in Paris. In fact Holsti lists thirty inter-state wars between 1918 and 1941, at least eight of them arising directly from the post-1918 settlements. As he also notes, the number of states in the international system almost doubled, from twenty-two in 1914 to forty-one in 1918, and for the first time all continents would be represented in the new international body—the League of Nations. Far from creating a more tranquil world, the years 1918–41 would prove to be the most violent in international relations since the era of the French Revolution and Napoleon. Territory remained the single most important source of international conflict, followed by state or regime survival, and treaty enforcement. 'Most of the armed and diplomatic conflicts of the 1920s in Europe were contests over precise definitions of boundaries and adjustments to the territorial boundaries drafted at the Paris Conference.'[35]

Despite President's Wilson's rhetoric, the victors' freedom to fashion a 'new world' was severely limited. Moreover, despite being a historian and political scientist, Wilson had no great knowledge of European problems. The peacemakers were not only influenced and restricted by their own domestic opinion, but also in critical areas, such as parts of the former Habsburg Empire and Poland, they were confronted by *faits accomplis* in the form of new states established by force of arms in the closing weeks of the war and fortified by fervent nationalist enthusiasm. Furthermore, the statesmen assembled at Versailles pursued at least four contradictory aims: to enhance the victors' security, in most cases involving territorial gains; to punish the aggressors by exacting indemnities to compensate for the loss of lives, injuries, and physical destruction; to promote national self-determination in the creation of new states; and to implement the secret treaties involving Italy and the Middle East. Above and beyond all these, was the Wilsonian aim of creating an international body to preserve the peace settlements and prevent future conflicts.

The Treaty of Versailles embodied five main concerns in relation to the principal defeated enemy, Germany. These were the transfer of all Germany's former colonies to the victor powers under the euphemistic term of

1. The Battle of Königgrätz, 3 July 1866. The Prussian infantry's 'needle-gun' gave them an important tactical advantage in their decisive victory over the Austrians.

2. The Battle of Sedan, 1 September 1870. This is one of the earliest photographs to show troops in action. It was, and remains, difficult for the camera to capture the dynamism of combat.

3. British dead at Spion Kop, 24 January 1900. Although Britain eventually won the Boer War (1899–1902), her soldiers paid a high price for the initial ineptitude of the generals. More than 200 died in this 'acre of massacre'.

4. German infantry on the march to the Marne in August–September 1914. Such a casual, mass advance would have been unthinkable after the trench deadlock had set in at the end of the year.

5. 'The Conquerors'. Eric Kennington's painting of the 16th (Canadian Scottish)
Battalion towards the end of the First World War includes two Canadian Indians. To
the Germans the kilted Highlanders had a fearsome reputation and were known as
'ladies from Hell'.

6. Armistice Day, 1918. Peace descended on the Western Front on the eleventh hour of the eleventh day of the eleventh month. There was widespread rejoicing on the home fronts, but many combatants experienced a sense of anti-climax.

7. The Allied Commanders at Metz, 8 December 1918. Pétain (foreground) receives his Marshal's baton. Left to right: Joffre, Foch, Haig (behind him Foch's chief of staff Weygand), Pershing, Gillain (Belgium), Albricci (Italy) and Haller (Poland).

8. Tyne Cot Cemetery near Ypres. This notorious sector was the scene of almost ceaseless attrition for the British Expeditionary Force from October 1914 onwards. The prominent Cross of Sacrifice stands over a German bunker and, with nearly 12,000 burials, this is one of the largest and most poignant centres of pilgrimage on the Western Front.

9. Omaha Beach, Normandy, 6 June 1944. Although the Americans suffered heavy casualties in establishing a bridgehead, Robert Capa's photograph shows that order was quickly imposed and a purposeful advance begun.

10. Americans march in triumph through the Champs Élysées in September 1944, a month after the liberation of Paris. This photograph deserves to be better-known than that of the ephemeral German victory march through the French capital in 1940.

11. Russian soldiers raise the Red Flag over the Reichstag in Berlin in May 1945.
This photograph retains some of its potency as a symbol of heroic victory over
Nazism, but also has ironic undertones in view of the Soviet Empire's more recent
collapse.

12. Fifty-two German U-boats at their mooring in Linshelly harbour, Northern Ireland, after the German surrender in May 1945. Few photographs convey a more dramatic image of the consequences of defeat in the war at sea.

13. Aftermath of the dropping of the first atomic bomb at Hiroshima on 6 August 1945. This cataclysmic event, and the similar fate of Nagasaki three days later, forced the Japanese Emperor and government to accept the reality of defeat.

14. An Israeli armoured column advances into Egypt during the Six-Day War. This photograph captures the euphoria of military victory.

15. Egyptian prisoners captured during the Israeli advance, 9 June 1967. This remarkable photograph reveals the spirit of demoralization which usually pervades a defeated army.

16. The Argentinian warship *Belgrano* is sunk during the Falklands War, 1982. Its captain, and some British critics, claimed that it was heading away from the war zone, but for most students of the war it was a legitimate target, and its loss significantly affected Argentinian morale.

17. The ending of the Gulf War: destruction on the Basra Road, Kuwait, 1991.

mandates; the establishment of the League of Nations; the surrender of all territory annexed by Germany; the disarmament and occupation of (parts of) Germany to prevent her from waging war in the future; and, finally, the payment of reparations.

It is not necessary to discuss all these points in detail. As regards the territory forfeited, Germany gave up Polish lands occupied in the 1790s, north Schleswig annexed in 1864, Alsace-Lorraine taken from France in 1871, and Russian territory occupied under the terms of Brest-Litovsk. Of all these cessions of territory, those affected by the revived state of Poland caused by far the most resentment in Germany, partly on strategic grounds, in that the 'Polish corridor' isolated East Germany from the parent state, while Danzig was declared a free city under the League of Nations control; and partly on ethnic grounds, in that some two million Germans (and more German speakers) were included in Poland. However controversial, this part of the treaty did not contravene any of Wilson's fourteen points.[36]

The enforced disarmament of Germany, though never completely implemented, was effective for about a decade in removing the threat of renewed aggression against her neighbours. But since the severity of the terms (no warships, no tanks or heavy artillery, no military aircraft, no general staff, and an army limited to 100,000 long-service volunteers) were demeaning to an undoubted great power, there was every incentive to evade them. This is what happened in some crucial respects, such as the clandestine production of guns and of aircraft adaptable to military functions. The failure of other powers to disarm to anywhere near the same extent, as the peacemakers in 1919 piously assumed would happen, only served to give greater urgency to German demands for parity in the early 1930s. Finally, the victors unwittingly did Germany a favour by obliging her to concentrate on high-quality personnel and training, modern weapons and equipment, and mobility in theory and practice.

The article (231) making Germany reponsible for all loss and damage sustained by the Western powers, 'as a consequence of the war imposed on them by Germany and her allies', was later to prove a potent weapon in German propaganda, abetted by the guilty consciences of British liberals, much encouraged by J. M. Keynes's polemic *The Economic Consequences of the Peace* (1919). But in reality this punitive aspect occupied only a limited place in the Treaty, and its severe demands (eventually set at £6,600 million plus interest) were mitigated both by practical problems and by

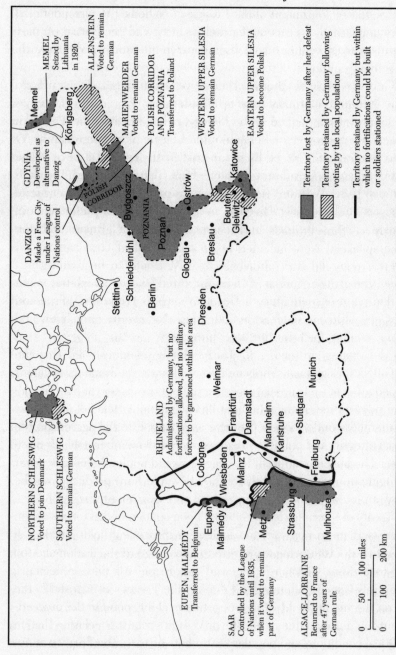

MEMEL
Seized by
Lithuania
in 1920

ALLENSTEIN
Voted to remain
German

MARIENWERDER
Voted to remain German

POLISH CORRIDOR
AND POZNANIA
Transferred to Poland

WESTERN UPPER SILESIA
Voted to remain German

EASTERN UPPER SILESIA
Voted to become Polish

GDYNIA
Developed as
alternative to
Danzig

DANZIG
Made a Free City
under League of
Nations control

POLISH CORRIDOR

POZNANIA

Memel

Königsberg

Bydgoszcz

Ostrów

Beuthen
Gleiwitz

Katowice

Schneidemühl

Poznań

Stettin

Glogau

Breslau

Berlin

RHINELAND
Administered by Germany, but no
fortifications allowed, and no military
forces to be garrisoned within the area

Dresden

Weimar

Munich

Frankfurt
Darmstadt

Mannheim

Stuttgart

Cologne

Mainz

Wiesbaden

Karlsruhe

Freiburg

NORTHERN SCHLESWIG
Voted to join Denmark

SOUTHERN SCHLESWIG
Voted to remain German

Eupen

Malmédy

Metz

Strassburg

Mulhouse

EUPEN, MALMÉDY
Transferred to Belgium

SAAR
Controlled by the League
of Nations until 1935,
when it voted to remain
part of Germany

ALSACE-LORRAINE
Returned to France
after 47 years of
German rule

Territory lost by Germany after her defeat

Territory retained by Germany following
voting by the local population

Territory retained by Germany, but within
which no fortifications could be built
or soldiers stationed

0   50   100 miles
0   100   200 km

Map 10. German territorial losses after the First World War

considerations of Germany's ability to pay.[37] Whether it was politic to impose *any* financial and economic penalties in view of the resentment they would provoke may be questioned, but the much-attacked 'war guilt' phrase (never actually employed in the Treaty) appears to be a reasonable description of Germany's role in 1914, whether viewed from the standpoint of 1919 or in the light of modern historical scholarship.

The failure of the United States Senate either to ratify the Treaty of Versailles or to approve membership of the League of Nations dealt heavy blows to the credibility of both, and in particular deprived France of the safeguards against a German revival on which Clemenceau and Foch had set so much store. With or without American participation, the League of Nations was fundamentally flawed as an institution for the preservation of international order. Its three main supporters, France, Britain, and the United States, held irreconcilable views of its role in the face of 'aggression': the French regarded it essentially as an instrument to be used against Germany; the British Foreign Office supported its role as arbitrator but would deprive it of teeth; and the idealists such as Woodrow Wilson and Robert Cecil wished to see it act in both arbitration *and* coercion but would give it no standing force for the latter purpose.[38]

In a devastating critique,[39] Sir Harry Hinsley demonstrated that the League of Nations was not simply unlucky in the way international relations developed after 1919: from its very conception it was bound to fail. To take just one point of its salient statement of obligation, Article 10; under this article members undertook to preserve as against external aggression the territorial integrity and political independence of all member states. If these obligations had been honoured the League 'would not have eliminated war and struggle in international relations . . . It would have produced, on the contrary, an intensification of struggle and an extension of war.'

The Treaty of Versailles has frequently been condemned as too severe on Germany and more rarely as too lenient; a third view holds that it fell between two stools in being too severe to encourage reconciliation and too lenient to prevent Germany's renewed aggression. All these reflections, however, rely heavily on hindsight. Certainly the Treaty was imposed by the victors on the vanquished (without negotiation at any point in the proceedings), but this author agrees with Trevor Wilson's robust conclusion that, in the political conditions and popular sentiments of 1919, the Treaty was not excessively severe.[40] True, Germany suffered some permanent territorial losses and the temporary loss of the Saar and the (military use of) the Rhineland, but the victors' occupation was limited in time and area, and she remained a powerful, ethnically unified industrial state. The military and naval restrictions *were* severe and humiliating but would inevitably have been revised in time. In fact revision of these clauses (and the flagrant breach of other aspects of the Treaty) were to occur in extreme political conditions in the 1930s unimaginable to the peacemakers in 1919. The Treaty did not *necessarily* fail on account of its contents; it failed because the circumstances in which it was produced rapidly ceased to obtain. Of the four main belligerents against Germany (though only three of them victors), Russia after 1918 was engulfed in revolution and in any case excluded from international institutions; the United States withdrew from European commitments and would play only a marginal role in the post-war fate of Germany; and both Britain and France emerged seriously weakened by their war efforts. If Britain and France had single-mindedly preserved and developed their wartime collaboration against Germany, they might just have been strong enough together to prevent the latter's renewed aggression, but they proved to be anything but united in the inter-war period. Here, then, was a vivid illustration of the ironies and limitations of military victory. The victory itself had been worth winning and impressive, but the issues were too complicated and the means lacking to convert military success into an enduring peace settlement. Which statesman in 1919 was nearer the truth in his understanding of human nature and the behaviour of sovereign states, Clemenceau or Woodrow Wilson? It is a dilemma as to what is to be expected from international co-operation which, as the Bosnian civil war demonstrates, we are no nearer answering in the 1990s.

Whereas many front-line combatants expressed low-key reactions to the war's sudden ending, including surprise, puzzlement, disappointment, and, above all, relief to be still alive, it was mainly civilians on the home front who shouted 'Victory' and indulged in an immediate orgy of unrestrained celebration.[41] In Paris, as a socialist writer noted in his diary, the rejoicing did not even wait for the guns to boom at eleven on 11 November: 'Bells are ringing. The air is full of their peals. Soldiers dance with ecstasy. It is a pleasure to witness their delight. Tragedy was looming over them.' When the 'peace barrage' sounded people in the streets sang, wept, and shouted for joy. Armfuls of flags were carried outdoors. Captured

German cannon were dragged in triumph along the boulevards. Solid masses of people, locked arm-in-arm, were singing and 'Vive-la'-ing everything but the Boche.

In the evening Lady Scott, widow of the explorer, watched delirious crowds playing 'kiss in the ring': 'mad, wild scenes, girls dressed as widows were dancing with the rest.' In Belfort a wounded French sergeant was 'carried away by the popular rejoicing as by a heady wine. I fancied myself bearing on my shoulders the awesome glory of the victors, the triumphant glory of those who survived.'

Similar spontaneous outbursts of relief and rejoicing took place in London, New York, and throughout the towns and villages of the victorious nations. Margot Asquith found the atmosphere in London 'more like a foreign carnival', with huge flags flying everywhere and pedestrians dancing on the pavements. Though saddened by the death of his son Raymond on the Somme, Herbert Asquith was heartened by the thought that 'war would now be recognised as an absolute anachronism'. Huge crowds besieged Buckingham Palace chanting 'We want the King' until the Royal Family appeared on the central balcony, which was draped in red and gold. Despite a steady drizzle the crowds remained for hours singing 'keep the Home Fires Burning' and other popular songs. On a slow, 9-mile drive through London in open carriages George V was touched by the fervour of the cheering crowds. The day ended in 'a storm of delirious thanksgiving'. As Osbert Sitwell noted sardonically, nothing like this had been seen since 4 August 1914 when the crowd was unwittingly 'cheering for its own death'. The revelry in Trafalgar Square went on without let up for three days. All too soon, however, the harsh realities of peacetime would intrude and with them the sadness for the millions who would never return.

Sir Douglas Haig did not return to England until 19 November 1918. He was welcomed officially at Dover and Charing Cross, but what touched him most was the wholehearted enthusiasm of the crowds which packed his route to Buckingham Palace. 'I could not help feeling how the cheering from great masses of all classes came from their hearts.' When he returned home to Kingston in the evening he was welcomed by a crowd which he estimated at 10,000, 'with torches and three bands, mostly working men and women from the Sopwith Aeroplane Works'. He felt that the spontaneous welcome he had received ever since landing showed that the people appreciated what had been accomplished by the Army under his leadership.[42]

Official celebrations were organized in the capitals of the victorious powers, including a grand parade in Paris on Bastille Day, 14 July 1919; and five days later in London a parade of 18,000 troops was headed by a galaxy of Allied military leaders, including Foch, Haig, Robertson, Pershing, and Jellicoe. But already the euphoria of November 1918 was fading and there was little evidence of what would now be termed 'triumphalism'. Wartime rhetoric and exaggerated expectations were rapidly giving way to the disappointments of the harsh post-war world. For example, demobilization had been ineptly handled in Britain and France, and even when the process was made fairer and speeded up, many veterans found they had no job to return to. Commemorating the dead became big business as war cemeteries were prepared all along the Western Front, and war memorials erected in villages and churches throughout Britain. 'Between 1920 and 1923 British shipments of headstones to France reached 4,000 a week.' On 11 November 1920, after a complex procedure to ensure anonymity, the unknown soldier was brought back from France and buried in Westminster Abbey. Within two days more than 100,000 wreaths had been laid at Sir Edward Lutyens's newly erected cenotaph in Whitehall. In Paris it was noted that attendance at the annual celebration of victory was already declining in 1920 as the public sought amusement and escape from the war.[43]

Not surprisingly, there was soon critical speculation about the true meaning and purpose of the war in which some 60 million troops had been mobilized (excluding the United States and Japan), of whom over half became casualties, with more than 8 million killed, 21 million wounded and nearly 8 million taken prisoner.[44] It had been less than edifying to see the victors' representatives quarrelling openly over the peace terms which, even among the victorious nations, left millions dissatisfied. Civil strife, in some cases escalating to full-scale civil war, raged in Russia, Germany, Poland, Hungary, Ireland, and Italy. Turkey and Greece fought fiecely over the terms of the Treaty of Sèvres, which were revised at Lausanne in 1923 in recognition of the former's military successes. Fear of Bolshevism pervaded Western societies.

Historians of literature and culture stress the sharp break between wartime experience and the post-war world. Suddenly, the war passed into history: painful to remember for the survivors, difficult to discuss with non-combatants, and almost impossible to write about objectively.[45] With very few exceptions, war recollections were consciously repressed or filtered through assumptions and beliefs developed in the very different post-war world.

In view of the many profound problems of adaptation to a new and drastically changed world, nominally at peace, we must be wary of generalizing about the alleged mood of disenchantment and disillusionment, and particularly of reading these attributes back into the war experience itself. Many front-line combatants, like R. H. Tawney, a socialist and later a distinguished economic historian, retained the idealism for which they had enlisted because only if those ideals were valid would their suffering and the loss of comrades be justified; it was all too often the patriots, politicians, and profiteers who abandoned their ideals—if they had ever had them. Hence the justified anger in many soldiers' wartime letters and their subsequent sense of alienation from post-war civil life.[46]

Robert Wohl also makes the important point that the main characteristic of the best and most honest of the war literature is ambivalence. If some intellectuals, writers, and poets experienced the war in terms of horror, catastrophe, brutalization, and fear, it was also experienced by many of the same men—or men of similar outlook—as an opportunity, a privilege, and a revelation. 'The same men who cried out at the inhumanity of the war often confessed that they loved it with a passion and wondered if they would ever be able to free themselves from the front's magical spell.'[47] For thousands who experienced active service, the First World War was by far the most exciting and uplifting episode in their lives. Combat brought an enormous simplification of life and a paring down to bare essentials. It forced individuals to confront their own fundamental strengths and weaknesses. Some failed the test but many returned home feeling that they had become different and better men. Frederic Manning's justly celebrated work of fiction *The Middle Parts of Fortune* (originally published in a bowdlerized edition as *Her Privates We*) brilliantly exemplifies this personal challenge in the experience of its 'hero', Bourne.[48]

But, as Wohl carefully concludes, no unitary pattern of response emerges. The front experience weakened some and strengthened others; drove some to the right and others to the left. His point is that this extremely complex and ambivalent experience of war affected a specific *generation of young men* and invested the recurring sense of a 'generation gap' with a unique emotional charge in the 1920s and 1930s.

All this may seem far removed from the notion of victory, but the point is that the 'real war' ceased to exist after November 1918 to most people except military historians. The cynicism and disillusionment now generally taken to characterize combatants in the later part of the war were—in so far as national moods can be characterized in this way—largely the products of the first few years of peace. It was precisely because idealistic expectations (fanned by political rhetoric and propaganda) had been so inflated during the war that the subsequent reaction in some individuals was so strong. Class barriers, if they had ever been removed, were resurrected; selfishness was not replaced by co-operation; suffering and sacrifice were not adequately compensated; and conflict within and between nations did not cease. By the end of the 1920s men and women of all age groups and in all parts of Europe suffered a tremendous sense of anticlimax. It was in this atmosphere at the end of the 1920s that the boom in 'anti-war' books flourished. Robert Graves expressed one reaction in the title of his memoirs, *Goodbye to All That*, while Erich Maria Remarque's *All Quiet on the Western Front* made an immediate and overwhelming impact on readers who had experienced this disillusionment. But, as Eksteins neatly put it, the latter best-seller was not 'the truth about war'. It was, first and foremost, 'the truth about Erich Maria Remarque in 1928'.[49]

By the late 1920s furthermore, there was much less public interest in how the First World War had been won—that positive aspect was now being downplayed if not actually derided—than in why it had broken out. In Britain especially, though not exclusively, the main problem seemed to be how a 'Second World War' might be prevented. Reliance on open diplomacy and the League of Nations might provide part of the answer, but in the view of radical and left-wing politicians (such as Arthur Ponsonby) public opinion had a vital role to play because it was naturally pacific. In *Falsehood in Wartime* (1928) Ponsonby sought to prove that in 1914 and after in every country the peace-inclined but ignorant peoples had been deliberately misled by bellicose propaganda. This extremely popular peace movement of the late 1920s drastically misrepresented actual public beliefs and behaviour in the First World War. Consequently 'It was only in the 1920's and 30's, with the rise of more potent imagery of peace and international co-operation, that the public required a new iconography of war—one which emphasized its futility and suffering.'[50]

It is commonplace to remark that after 1918 there was a widespread revulsion on the part of liberals and democrats against the utility of war (as

Bloch, Norman Angell, and others had vainly argued before 1914), and more especially against the positive view of war as an instrument of state policy. Jomini was at last toppled from his pedestal as the doyen of military theorists whose books had been used to mediate the principles of Napoleonic war to generations of army officers. Napoleonic principles seemed of doubtful relevance to officers who had experienced Verdun or Passchendaele, or in general to a new style of conflict between nations waged by aircraft, tanks, and submarines. But Jomini at least had the saving grace of being a Swiss national who had served France and Russia, Britain's recent allies. No such plea of mitigation could be made for Clausewitz, a Prussian career professional from the age of 12. B. H. Liddell Hart, very influential as a military journalist with the *Daily Telegraph* (1925–35) and the *The Times* (1935–9) as well as being a prolific military historian, led the way in criticizing Clausewitz as the 'Mahdi of Mass' whose preoccupation with battle and mass armies had allegedly contributed significantly to the stalemate and slaughter of 1914–18.[51] These criticisms accorded well with the propaganda line that the war had been fought against 'Prussian militarism', and there were very few experts in the West at this time to put the case for Clausewitz's ideas about the nature of war.

As a soft alternative to the bleak Clausewitzian view of international relations as an endless power struggle, idealistic supporters of the League of Nations put their faith in the Covenant as a means of reducing or stopping inter-state conflict. Though soon to be discredited when serious challenges to its authority were mounted in 1931 and after, the League did enjoy some modest successes in dealing with limited conflicts in the 1920s.[52]

The apotheosis of internationalist idealism occurred in 1928 with the proclamation of the Kellogg–Briand Pact, whose sixty-five signatories (including Germany, the USA, and the Soviet Union) agreed to renounce war as an instrument of national policy—though reserving the right to defend national interests. The pact made no provision for the punishment of 'aggressors' and was soon perceived to be little more than the empty rhetoric of a brief period of optimism in international relations.

By the end of the 1920s it was becoming clear that the reaction against violence and war was far from universal. Fascist movements of the radical right did not merely accept war as a valid instrument of policy, they glorified it as the highest expression of nationalist ideology. At the other political extreme the revolutionary left, as represented by the Soviet Union, believed that both internecine and international conflict were necessary to usher in the millenium. Between them these extremist movements destroyed the fragile ideological consensus—already badly undermined by the war—on which the international system was based.[53]

Few military leaders were convinced that the post-war peace treaties and the League of Nations had significantly reduced the likelihood of inter-state conflict. This did not necessarily mean that they were 'militarists' in the sense of welcoming war for professional or ideological reasons. Their pessimism derived from two quite different sources. The first was that there were obvious threats to the security of the satisfied powers both in Europe and in their overseas empires. Behind the pious rhetoric of the League of Nations Covenant, the Locarno Treaty, and the Kellogg–Briand Pact, it soon became apparent that the responsibility of each sovereign state to protect itself and its interests remained unchanged by the 'war to end all wars' and the subsequent attempts to create a more peaceful and co-operative world order.

The second consideration lay in the military sphere. Bloch and other pre-1914 prophets of the indecisiveness of war due to the superiority of the means of defence had taken too polemical and static a position. The rapid development of motor vehicles, tanks, and aircraft had provided the means of mobility, but the commanders and staffs had also eventually evolved operational methods to overcome the trench deadlock. The superiority of the defenders, which looked so enduring to Bloch in 1900, was very much open to question by 1918. Military historians are still debating which were the most important innovations and who deserves credit for them, but there is no dispute that (at the latest) from March 1918 onwards mobility and manœuvre were restored to the campaign in Western Europe, and that after the German March offensive had narrowly failed the Allied forces displayed impressive tactical flexibility in advancing steadily to a definite victory.

It was not therefore surprising that while the victorious powers, especially France and Britain, paid lip-service to the most striking components of offensive, mobile warfare—tanks and tactical air power—it was in defeated and disarmed Germany that the seedlings of what was later to be called 'blitzkrieg' were most promptly and deeply rooted. Recent scholarship has stressed the great significance in this process of the regime of General Hans von Seeckt as head of the *Truppenamt* in the Ministry of War (as a substitute for the banned general staff) between 1919 and 1926.[54] Not only did von Seeckt and his colleagues evade many of the restrictions of the Versailles Treaty, notably by clandestine co-operation with the Soviet Union from 1922 onwards, but were also progressive in terms of doctrine and training in preparation for future war. Germany's weakness in numbers of troops and equipment, even *vis-à-vis* Poland, let alone France, put a premium on manœuvre. Consequently, as early as 1921 the Reichswehr was running motorized infantry exercises in the Harz Mountains using requisitioned civilian trucks. Under Seeckt's guidance the *Military Weekly* displayed a voracious and vicarious interest in international developments in motorization, mechanization, and the military application of air power. Whereas France's military leaders broadly opted for a defensive strategy, as represented by the Maginot Line, this lesson from the recent war was rejected by the Reichswehr. The latter believed that Germany was more vulnerable than its enemies to a war of attrition; another war of the trenches would produce stalemate at best and certain economic collapse in the long term. The Reichswehr therefore boldly adopted the doctrine of a rapid war of manœuvre, leading to an early decision entailing the annihilation of the enemy forces. The idea of striking first and quickly offered the only real chance of victory once the army was expanded and modern weapons acquired. It is no exaggeration to claim that most elements of the blitzkrieg of 1939 and 1940—though not the Panzer divisions—derived directly from the regulations issued by the Reichswehr in 1921 and 1923. Moreover, in Seeckt's manœuvres Reichswehr units learned to ignore the continuous front and to advance rapidly without regard to their flanks. Perhaps most important of all, the Reichswehr stressed the use of combined arms as its central tactical principle. Both at the officer-cadet and general-staff training levels Seeckt's regulations stressed the value of attachment to other arms and a complete familiarity and competence in the tactics and weaponry of all branches of the service.[55] This freedom from the parochialism of the single arm and the low intellectual horizon bounded by life in the regiment gave the *Wehrmacht* a conspicuous advantage throughout the Second World War.

The second, and broader, deduction from the experience of 1914–18 was that the next great conflict would be total, both in the sense that civilian and military potential must be mobilized to the maximum, but also in the more ominous sense that the enemy population would be a legitimate—perhaps indeed even the main—target rather than the traditional objective of the armed forces. Thus the proponents of total war, of whom Ludendorff was only the most extreme, envisaged an all-out struggle of propaganda and subversion, of starvation by blockade, and of remorseless bombing from the air. If these methods did not bring about the enemy's complete demoralization and surrender, he could be finished off by swift land operations spearheaded by armoured forces.[56]

Thus belief in the feasibility of decisive victory remained the guiding principle among military commanders and theorists despite the sobering attritional nature of the First World War. It is notable that even such a liberal and progressive thinker as Liddell Hart, who in so many respects viewed himself as a severe critic of the higher conduct of the First World War, nevertheless continued to believe that decisive victory was possible in the conditions of modern industrial war. Indeed, Liddell Hart continued to think in traditional terms of decisive generalship, as distilled by Jomini, whom he described as 'the pillar of "sound" strategy', provided it was supported by the adoption of modern infantry and armoured tactics. Whereas Liddell Hart was primarily interested in restoring decisiveness to war by finding alternative 'indirect approaches' to the earlier preoccupation with the decisive battle, J. F. C. Fuller continued to focus consistently on making the decisive battle once again the principal objective of the theory and practice of war.[57]

Thirdly, the aspect of modern 'total war' in which most theoretical faith and public anxiety was invested was the notion of a rapid decision through what came to be termed strategic bombing.[58] In the inter-war period it was generally assumed that the special terrors of gas would be added to the effects of high explosives and incendiaries. Though he was never predominantly an air theorist like the Italian Guilio Douhet or some RAF strategists, Liddell Hart, too, briefly subscribed to the paramount importance of strategic air power as a war winner. His flirtation with this concept is interesting both because he was predominantly an army-oriented writer with excellent service contacts, and because he regarded himself as a humane thinker dedicated to inculcating more acceptable alternatives to the bloodbath of 1914–18.

In *Paris or the Future of War* (1925) Liddell Hart's basic premiss was that the Allies had followed an erroneous strategy in the First World War; namely, the destruction of the enemy's armed forces in the main theatre of operations. Ordinary citizens had paid a terrible price for this blunder, 'yoked like dumb, driven oxen to the chariot of Mars'. The function of grand strategy was to discover and exploit the Achilles' heel of the enemy nation (just as Paris had exploited the Greek champion's weak spot—hence the punning title), which he saw as the modern state's vulnerability to a sudden and overwhelming blow from the air. 'A nation's nerve system . . . is now laid bare to attack and, like the human nerves, the progress of civilization has rendered it far more sensitive than in earlier and more primitive times.' He painted a terrifying picture based on a projection of the limited bombing experience of the First World War into the future:

Imagine for a moment that, of two civilized industrial nations at war, one possesses a superior air force, the other a superior army. Provided that the blow be sufficiently swift and powerful, there is no reason why within a few hours, or at most days from the commencement of hostilities, the nerve system of the country inferior in air power should not be paralysed.

Such an air bombardment would be terrible but it would be over quickly: (British) aircraft would be able to reach and destroy cities such as Essen or Berlin *in a matter of hours*. Liddell Hart went so far as to assert that ultimately the air would become 'the sole medium of future warfare'.[59]

In the disappointing aftermath of the First World War, and particularly from the later 1920s, it became a matter of faith in much liberal and pacifist ideology that the First World War had been 'futile' and that resort to war in the future was scarcely conceivable, except as the ultimate measure in self-defence. But that supposed flagrant aggression in what was expected to be a more pacifically inclined world in which disputes could be resolved without conflict by the League of Nations. At worst 'sanctions' would bring an aggressor to reason. A dangerous gulf consequently opened between the 'idealists', who were very influential in some democratic countries, and the 'realists', who risked being labelled as warmongers if they called attention to deficiencies in the armed forces. This is, of course, to put complex issues in simplistic terms but it remains the fact that in Britain in particular it became unfashionable to regard the First World War as a conflict which had been worth entering and fighting to a victorious conclusion. Thus 'imperial defence' and 'home defence' were acceptable notions, but a 'continental commitment' to resist a resurgent Germay was not.[60]

Pride in the mighty achievement of winning the war was eroded too by the growing realization that the defeat of Germany, and some other aspects of the peace settlement, would not be permanent. It may be contended, however, that these failures and disappointments did not lie within the war itself or even necessarily with the errors of the peacemakers. The objective realities of power at the conclusion of hostilities determined the duration of the peace settlements.[61] A comparison of the power balance as between Germany and her opponents in 1918 and 1945 is instructive, particularly as regards the roles of the Soviet Union and the United States. After the First World War, in short, the anti-German coalition patently lacked the strength and resolution to prevent the former's rapid resurrection as a great power and the dominant force in Central Europe. In sharp contrast, after 1945 Germany was divided, and although the Soviet Union and the United States were at loggerheads, both had an interest in preventing a speedy German military revival.

Few statesmen or scholars between the wars really pursued the questions of the likely consequences had Britain not participated in the war or not emerged on the winning side. In bemoaning the failure to achieve the nebulous aim of crushing German militarism and the utopian goal of ending war itself, critics underplayed the positive achievement of securing the more modest original aims such as restoring Belgian independence, and removing the German land threat to the Low Countries and the Channel coast. The destruction of the German high seas fleet, albeit self-inflicted as a final act of defiance, was a remarkable achievement scarcely conceivable in 1914. In addition France had recovered Alsace-Lorraine and for the foreseeable future, with the Rhineland under Allied occupation, was secure against a sudden German offensive. Those who argued that victory was indistinguishable from defeat were guilty of muddled thinking: what they really meant was that victory had been achieved at too great a cost. That was a tenable viewpoint in hindsight but difficult to apply while the war was being fought (at what point should Britain have withdrawn to avoid further losses?), and impossible to assess afterwards. The implications for the West of the harsh terms of Brest-Litovsk, or the more recent revelations of German war aims, have seldom been taken into account by those who have asserted that the First World War was not worth winning. Even to have settled for 'a draw' on the Western Front would have had dangerous conse-quences long before 1939.

In a curious way, coming to terms with the ultimate justification of the British role in the First World War has been made more difficult by partici-pation in the Second. The Kaiser was manifestly not such an evil force as

Hitler, and 'Hunnish' atrocities in the First World War, exaggerated by propaganda, were on a lower plane altogether than the systematic application of Nazi racism and barbarism in the Second. Consequently, whereas even the radical, anti-military historian A. J. P. Taylor was prepared to concede that the Second World War was ultimately, 'a good war', he adopted a supercilious tone towards the statesmen and military leaders who had waged the meaningless war of 1914–18.

But if the attritional nature and very heavy casualties in the latter render inappropriate the description of 'a good war', was it not a 'necessary war' caused by the German drive for hegemony and fought, particularly in the final phase after Russia's collapse, to defend liberal democracy in Western Europe and North America against an expansionist military autocracy? It is surely relevant to note that despite the profound mood of disenchantment in some quarters about the outcome of the First World War, and the increasingly severe criticisms directed against both civil and military leadership in it from the later 1920s, a new generation was prepared to make similar sacrifices in 1939 to stop German aggression a second time, and that before anywhere near the full horrors of Nazi conquest had been publicized. While it may now seem excessively rhetorical to describe the First World War as 'one of freedom's battles' it is appropriate to reconsider the war's purposes, costs, and outcome from the viewpoint of the generation which made unprecedented sacrifices in order to secure victory.[62] Whether or not the advantages gained by military victory were properly exploited after 1918 is a separate issue.

Despite their adherence to international declarations outlawing the waging of war, and the widely held belief that the First World War had been so costly and unproductive that it must never again be repeated, the satisfied major powers such as France, Britain and her Dominions, and the United States, as well as smaller powers such as the Low Countries, Czechoslovakia, Denmark, Norway, and even Switzerland, gradually came to realize in the 1930s that they might be forced to fight in order to defend their homelands or national interests. Whether war for such unmilitaristic states would constitute 'an instrument of policy' or rather be regarded as a tragic *failure* of policy, once forced into hostilities they would need to seek victory as in the past.

The Pursuit of Victory in the
Second World War

Despite the profound and widespread reaction against war after 1918, exemplified in the 1920s by the disarmament conferences and the proclamation of the Kellogg Pact, there was by no means universal acceptance that war had ceased to be a viable instrument of state policy. In eastern Europe, for example, the Bolsheviks triumphed in the civil war and established their authority over the bulk of the former Tsarist empire; while the Poles decisively defeated a Russian invasion at the battle of Warsaw in 1920, and enhanced their victory by extending their frontier eastward at the Treaty of Riga the following year. Thus, while in Britain and France, 'men counted their dead and meditated on the futility of war, the peoples of eastern Europe counted their gains and losses and observed the efficacy of force'.[1]

The term 'Second World War' is, of course, a historical construct which signifies very varied chronologies, myths, and legacies in different parts of the world. Some contemporaries placed the European conflict in the context of a 'thirty years war' (1914–45) concerned with the unfinished business of Germany's bid for hegemony, which had only received a temporary check in the Treaty of Versailles. From a narrower, or more precise, viewpoint the conflicts between September 1939 and June 1941 constituted a European civil war ('The Last' in John Lukacs's optimistic title),[2] with a subplot of local wars in the Baltic and the Balkans. Lastly, the Japanese had been at war in Manchuria since 1931, extended the conflict to the Chinese heartland in 1937, and initiated a truly global struggle by the attack on Pearl Harbor in December 1941. Hitler immediately linked the European and Pacific wars by recklessly declaring war on the United States, but even thereafter several anomalies remained. British and American co-operation with their Soviet ally was severely limited by the latter's ideological mistrust, but even so was far more substantial than the links between Germany and Japan. In particular Japan, though a member of the Axis, was not at war with the Soviet Union until August 1945. From the moment of her entry, the United States was in effect waging two great wars on opposite sides of the globe on the fundamental and enduring strategic assumption that Germany was the major opponent whose prior defeat would inexorably lead to that of Japan. This eventually proved to be an accurate assessment, but the terminal date of the Pacific war was still very hard to calculate in the early months of 1945, otherwise Stalin would not have been encouraged at Yalta to make war on Japan.

The inter-war advocates of mechanization, armoured warfare, and what has popularly become known as blitzkrieg were correct in their confident predictions that new means of mobility, protection, and striking power, if successfully combined in organization, doctrine, training, and deployment, offered the prospect of 'a revolution in warfare', that is of winning quick, decisive, inexpensive victories with technical means that had not been available in 1914. Further ingredients were necessary in the ideal recipe for victory, including the preponderance of mechanized forces and superior air power on one side; the waging of aggressive mobile operations against a more static and defence-oriented opponent; and, not least important, the readiness of the war leader to strike ruthlessly at the moment of his own choosing in defiance of such handicaps as international law, treaties, neutrality, and the civilities of diplomacy. The leaders of Italy, Germany, and Japan were to display these aggressive qualities from 1939 onwards abetted in the latter two at any rate by superb military instruments, to achieve the rapid and devastating annihilation of their enemies in a style unimaginable in the First World War. However, as Geoffrey Blainey has reminded us, it is a fallacy to believe that mechanization, in itself, will necessarily produce short, decisive wars.[3] What is essential, in addition, is firm political control for specific limited goals: in short, statesmanship of Bismarckian quality able to keep a grip of unfolding military events and to translate battlefield successes into peace treaties acceptable not only to the defeated but also to other important powers with an interest in the outcome. This implies that the political goals should also be precise, limited, and clearly thought out. Such statesmanship was conspicuously lacking in the aggressor states of the Second World War; indeed, one could go further and suggest that the very qualities and beliefs which made war and conquest seem attractive options were likely to rule out the statesmanlike virtues stipulated by Clausewitz and practised by Bismarck.

The Second World War posed in acute form the dilemma already apparent in earlier conflicts: the difficulty of terminating hostilities when the leaders of the nation or nations which are clearly losing and have no reasonable prospects of reversing the impending disaster nevertheless refuse to recognize their fate and determine to fight 'to a finish', entailing the destruction of their country and extreme suffering for their people.

The Second World War, accepting the conventional chronology, was notoriously one of two halves in the first of which (1939 to mid-1942) the Axis Powers won a series of dramatic victories which left them in command of a great deal of Europe, the Mediterranean and the Pacific. The peak of the Axis Powers' successes was reached at different times: Italy's in 1940, Germany's at the end of 1941, and Japan's in mid-1942; but by the beginning of 1943 the tide had turned decisively against all three. Industrial strength, superior manpower and indigenous resources, and maritime access to scarce materials (as well as superior political and intelligence systems), all seemed to make eventual Allied victory certain, provided only that the most powerful members continued to co-operate in pursuit of the same broad goals. Stalin's deep mistrust of his capitalist allies was well-known to Hitler, but the latter's reliance on a complete rift among his enemies—comparable to that which had saved Frederick the Great in 1762—persisted long beyond any rational calculation. Roosevelt's death in April 1945 came far too late (even assuming a drastic reversal of American policy was possible) to save Germany from being overrun from East and West. The point to be stressed here is that fanatical resistance, inspired in their nations by Germany's and Japan's leaders, unnecessarily prolonged the war by many months and should be held to blame for much of the destruction and suffering which their peoples experienced in 1944 and 1945—quite apart from the incalculable misery and casualties experienced in the vast territories still under their direct or indirect rule.

The United States injected a spirit of idealism into the Western objectives in the conflict at a much earlier stage than in the First World War.[4] Military victory, from the American and British viewpoints, would be largely barren if it did not yield a better peace dividend than the League of Nations. The new world-wide concert of liberal, democratic powers, to be known as the United Nations, would be explicitly a league to enforce peace. Germany (and by implication Japan also) must be punished and subjected to a long purgation involving occupation and re-education before they could be admitted to this club of civilized states. In August 1941 the still neutral United States and Britain signed the Atlantic Charter, which enjoined 'respect (for) the right of all peoples to choose the form of government under which they will live'; the signatories would recognize no territorial changes which did not accord with the freely expressed wishes of the peoples concerned; and they envisaged a peace which would afford to all nations 'the means of dwelling in safety within their own boundaries'. Further ringing declarations in favour of independence, political and religious freedom, and human rights were issued by the combined United Nations on 1 January 1944 and at Yalta in February 1945. When the United Nations Organisation was formally established after the end of the European war its members pledged 'to take effective collective measures for the prevention and removal of threats to the peace and for the suppression of acts of aggression or other breaches of the peace'.

Cynics have argued that this idealism did not filter down to ordinary front-line soldiers for whom the war presented only the stark option of kill or be killed, while the leaders, especially Churchill, were accused of pursuing victory for its own sake.[5] But at least as far as international declarations of principle were concerned, the Western governments had made it clear that the war was being fought for ideals. Victory must signal an end to the Hobbesian anarchy of sovereign states 'exacerbated by an arms race and tempered only by a balance of power'.[6]

Although the assignment of precise responsibility for the outbreak of the Second World War has been the subject of much historical debate, the deeper origins or causes of the war have provoked much less controversy than those of 1914. The reasons for the contrast are not hard to find. Whereas the onset of the First World War, for apparently frivolous reasons, shocked civilized Europeans grown accustomed to peace, the Second seemed a natural outcome of Fascist philosophy and organization which harnessed populist mass opinion to a militaristic and nihilist outlook. Nazi Germany, and to only a slightly lesser extent, Fascist Italy, professed ideologies which glorified war and set objectives which could only be achieved by it. In sharp contrast, in France and Britain, official and public opinion alike were strongly opposed to war: it was certainly not an acceptable 'instrument of policy' and should be resorted to only as a last resort. Consequently, in the strategic crises of the 1930s, neither thought in terms of opportunities to be seized, but only of calamities to be avoided by appeasement.[7]

Albeit from differing ideological standpoints and differing styles and objectives, the regimes of Italy, Germany, and the Soviet Union all regarded military force as a weapon to be used without inhibitions as an instrument of policy. Thus Mussolini waged a war of colonial conquest in Abyssinia and entertained further gargantuan ambitions of territorial expansion in north Africa, south-eastern France, and across the Adriatic. The Red Army fought an important campaign against the Japanese in July and August 1938 round Lake Khasan, and a much larger battle between May and September 1939 near the border with Manchuria and outer Mongolia. After initial reverses and heavy losses the Soviet forces triumphed and drove the Japanese back across the frontier.[8] In September 1939 Soviet forces occupied eastern Poland without serious opposition, and at the end of November Finland was attacked for the traditional purpose of territorial gain.

Between 1933 and 1939 Hitler enforced rearmament in Germany at breakneck speed. In the longer term, serious weaknesses would be revealed but by 1939 the results were sensational: in land and air power Germany had advanced in six years from being among the weakest states in Europe to being the strongest. Even whilst rearmament was far from complete Hitler was willing to risk war in successive crises from the Rhineland to the Sudetenland. In September 1938 he expressed frustration that his opponents had capitulated at Munich, thereby thwarting him of war.

Rearmament need not necessarily have forced Germany to go to war, but Hitler made abundantly clear to his service leaders that that was its purpose. In the short term (1939–41) the German economy and armed forces were prepared, in Philip Bell's apt phrase, 'for the sprint rather than the marathon'.[9] Though Hitler improvised in the timing and pretexts for attacking, there can be no doubt that the conquests of Poland, the Low Countries, parts of Scandinavia, and France resulted from his decisions. He also intended to nullify British resistance before turning against the Soviet Union where he had always entertained vast ambitions for territorial expansion. Whether Hitler intended to continue the pattern of short, decisive blitzkrieg campaigns indefinitely, or rather was preparing Germany for a protracted 'total war' in the East at a later date is still disputed. But this does not affect the point being made here; namely that Hitler and his service leaders employed war (and the threat of war) most effectively as an instrument of policy in 1939–41 to establish German dominance over the greater part of Europe between the Vistula and the Pyrenees and from northern Norway through the Baltic and Central Europe to the Danube basin and the Aegean.

To the generation of senior commanders of 1940, most of whom had experienced the First World War as middle-rank officers, the speed and decisiveness of the campaigns of 1939–41 were truly astonishing. It seems doubtful if even the theoretical proponents of blitzkrieg such as J. F. C. Fuller and B. H. Liddell Hart fully appreciated what could be achieved by the combination of armoured divisions assisted by tactical air power; indeed, Liddell Hart especially had emphasized the superior strength of the defensive in his publications in the later 1930s.[10] The French high command definitely held the view that operations would unfold at approximately the same tempo as in 1914–18. More surprisingly, with a few notable exceptions such as Guderian and Rommel (corps and divisional commanders respectively in 1940), the German high command (and Hitler himself) were also surprised and disturbed by the rapidity of their advance into France in 1940, fearing that the gap between the panzers and the support units would lay them open to a devastating counter-attack. Poland had been overrun more rapidly than Anglo-French assessments had expected, but then a Soviet 'stab in the back' had been added to the near-insoluble problems of defending her western frontier. The brevity of the resistance offered by Denmark and the Netherlands was not surprising, nor was the inability of France and Britain to bring effective support to Belgium when invited to do so only after the German invasion had begun. However, the *Wehrmacht*'s brilliant combined operations to seize Norway under the nose of the vastly superior Royal Navy was a remarkable feat of arms. By far the most astonishing example of blitzkrieg, which shattered virtually all strategic assumptions and future planning, was the overrunning and defeat of France in a mere six weeks at negligible cost.

Historians have also demolished the consoling myth that the German victories were due to overwhelming superiority in manpower and weapons.[11] In fact, in terms of divisions, artillery, and tanks, the opposing forces in the West in 1940 were approximately equal, and only in the important dimension of tactical air power was the *Luftwaffe* markedly superior in both numbers and operational role. In view of the widely accepted assumption that the attacker needs at least a three-to-one superiority, and also more recent research exposing faults and deficiencies in the *Wehrmacht*, the completeness of the German victories between 1939 and 1941 have come to seem less inevitable than they did at the time.[12] The main ingredients of victory are none the less apparent: an unrivalled combination of mechanized and air forces; well-trained troops who had learnt from recent experience; bold, offensive plans and ruthless execution. These advantages were quite sufficient to prevail over opponents who were less highly motivated and more defensively minded. Their high commands suffered the paralysis which theorists such as J. F. C. Fuller had predicted. Above all, Germany's early opponents were politically disunited and unable to trade space for time while they adjusted to the new, relentless style of warfare.

The rapidity and completeness of the German victories up to 1941 meant that their opponents' societies did not have time to become fully involved. The *Wehrmacht* defeated the enemies' armed forces and in every case a harsh peace settlement was imposed—and accepted. Lorenzo M. Crowell therefore errs, relying on hindsight, in citing these campaigns as examples of the illusion that 'with the correct tactical technique a decisive Napoleonic victory can be won'. The German generals *did* win a decisive victory in battle and it was not their fault that a lasting political settlement was not achieved.[13]

The terrible reality in June 1940 was that Nazi Germany dominated and effectively controlled Central and Western Europe and Scandinavia. Provided Hitler kept on good terms with the Soviet Union it was hard to see how this hegemony could be overthrown, but in fact his far-reaching aims in the East could only be achieved through conquest, and he lacked patience first to shore up his position in the West against what in the short term was largely token British resistance.

Britain was given an unexpected breathing space by Hitler's utter failure to plan ahead beyond the fall of France.[14] A more genuine conciliatory peace offer in May or June 1940 would not necessarily have gained immediate acceptance, but it would certainly have divided British political opinion, weakened the will to continue resistance in apparently hopeless circumstances, and possibly have resulted in the accession of a prime minister willing to negotiate.[15]

Alternatively, Hitler could have concentrated on the defeat of Britain by a combination of strategic bombing and naval blockade, rounded off, if

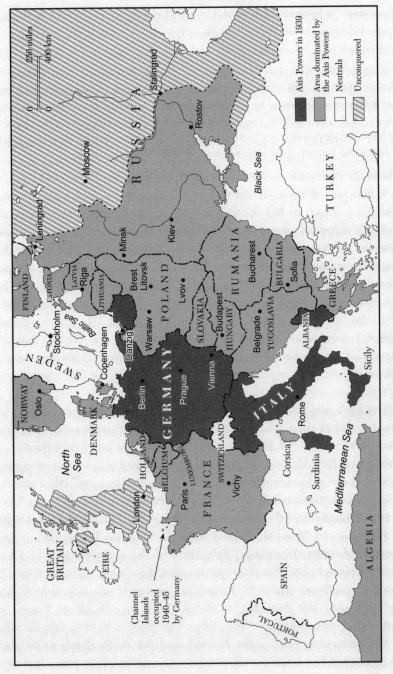

Map 11. German mastery of Europe, 1939–1942

necessary, by invasion. On Monday, 13 May Churchill asked the House of Commons for a vote of confidence in the new administration. His peroration included the following:

You ask, What is our policy? I will say: It is to wage war, by sea, land, and air, with all our might and with all the strength that God can give us: to wage war against a monstrous tyranny, never surpassed in the dark, lamentable catalogue of human crime. That is our policy. You ask, What is our aim? I can answer in one word: Victory—victory at all costs, victory in spite of all terror; victory, however long and hard the road may be; for without victory there is no survival.[16]

But in the following weeks and months, until indeed the end of 1941, the best hope for Britain was mere survival. Germany's pact with the Soviet Union, enhanced by her conquests, had rendered her virtually immune to Britain's favoured strategy of economic blockade; there was no realistic prospect of American entry into the European war; strategic bombing might boost home morale but was not an effective instrument for attacking Germany; and hopes of 'setting Europe ablaze' by encouraging anti-Nazi resistance movements proved to be largely illusory. For all his courage, optimism, and pugnacity, Churchill was aware that, objectively, Britain's prospects were weak: he could only hope that Hitler's failure as a statesman would throw away the near-impregnable position won by his own strategic risk-taking and the temporary superiority enjoyed by the German army and air force.[17]

The form and severity of German rule varied considerably throughout Europe. Other than a large part of Poland and a small area of Belgium, there was little formal annexation to the Reich, but other contiguous territories, including Alsace-Lorraine and parts of Slovenia, were so completely subjected to German needs that formal annexation seemed only a matter of time. Three other territories to the East were also completely subordinated to the Reich; namely the Government General of Poland, the Protectorate of Bohemia and Moravia, and the Baltic and Russian lands under a Reich Minister. Greece, Yugoslavia, Belgium, and the occupied area of France were all placed under military government, but in the last-mentioned the German plenipotentiary in Paris also exercised considerable indirect authority over the French government in Vichy. In effect France was partitioned until November 1942 when German forces occupied the whole of the Vichy-controlled zone. Norway and the Netherlands were placed under civilian Reich Commissioners but their parliaments, press, broadcasting, and police forces were either abolished or brought under strict German control. Denmark was the exception in being allowed to preserve a large degree of independence (and neutral status) with the King remaining in residence as as focus of loyalty for the government and the army. In 1943, however, the Germans took over the country, dissolved parliament, and disbanded the army.[18]

It is now widely recognized that Germany's early conquests were welcomed by substantial minorities in the conquered nations and that without a significant amount of active collaboration the Nazis' authoritarian regimes could not have been imposed so effectively. Although the full barbarity of Nazism was not displayed until after the invasion of the Soviet Union and the implementation of the Final Solution, the gratuitous violence against prisoners and civilians, the contempt for international law, and the sheer evil of the movement was immediately evident in the occupation of Poland and, to a lesser extent, in the West. Brutality and mercilessness were not the unfortunate by-products of military necessity; they were deliberately extolled in propaganda broadcasts and war films. Even before the harsh realities of the Nazi New Order became fully apparent, the first phase of the war witnessed the migration of peoples on a vast scale. Between 1939 and 1941 the German and Soviet regimes forced deportations of entire peoples amounting to about four million, with minimal provision for their resettlement.[19] Whether the racially favoured Nordic and Aryan nations would eventually have benefited and prospered in a victorious Greater Germany may be doubted, but in reality it is clear that military necessity caused an increasingly ruthless exploitation or despoliation of *all* occupied territories.

A recent study has demonstrated that, in defiance of liberal theory, the Nazis succeeded in making their conquests pay, especially as regards Central and Western Europe.[20] The initial German occupations were accompanied by special economic squads who seized what the Reich needed—in the way of raw materials, manufactured goods, currency, machinery, and rolling stock—leaving selected essential factories to be exploited as going concerns. At first economic exploitation was mitigated by German expectations of an early peace after victory in the East; but from the end of 1941 it became clear that Germany needed a greatly expanded arms base and must therefore exploit the occupied territories more ruthlessly, both for armaments and for other goods which her own war industry was no longer producing. In the West, those factories allowed to continue operating had to deliver a high proportion of their output to Germany. By August 1942 it was estimated that firms could only survive if at least three-quarters of their output was for the Germans. In the East only transport difficulties modified the ruthless transfer to Germany of all enterprises not absolutely essential for the maintenance of the workforce at the lowest level of subsistence. In general, industry in the East was 'limited to producing goods and services required by Germany's armed forces and administrators'.[21]

Similar controls were imposed on agriculture. Germany looked to western and south-eastern Europe for much of its food, and in the early part of the war kept these areas well supplied with machinery and fertilizers, but after 1942 failure to defeat the Soviet Union caused requisitions to be increased while the transport system came under greater strain. By the end of the war food production in occupied Europe had fallen by about a quarter and an ever-dwindling proportion was available to the civil population. Harsh rationing was introduced but its terms were often not honoured. Meanwhile between 1941 and 1943 requisitioned food supplies increased Germany's civilian ration by between a fifth and a quarter.

As the war went on the most valuable commodity to be exploited was labour. In the early war years some effort was made to attract genuine volunteers to work in Germany, but conditions were always disagreeable and deteriorated as a result of Allied bombing. Indirect coercion was applied in occupied countries and physical force was increasingly used. This was perhaps the single most important cause of resistance to German authority. In 1944 it was estimated that, out of five million foreign workers in the Reich, fewer than 200,000 had come voluntarily. By that year also some 40 per cent of the two million prisoners working in Germany as 'slave labour' were involved in arms production. Conditions for these workers, most of them captured on the Eastern Front, contravened all the rules of war and of human decency.[22]

There is, then, strong evidence that German coercion secured administrative collaboration, including financial and police co-operation, throughout most of occupied Europe, to permit economic exploitation at only modest costs to the occupiers. Small numbers of German troops, police, and administrators were needed to suppress overt signs of resistance. In late 1941, in France for example, half a million German soldiers, only a minority of whom were engaged in internal security, were able to pacify a nation of forty-two million. A recent study calculates that, through financial transfers alone, Germany was able to mobilize an annual average of 25 per cent of French economic potential, nearly 40 per cent of Dutch, Belgian, and Norwegian potential, and 70 per cent of Bohemian-Moravian potential over a four-year period.[23] Furthermore, despite continental raw material shortages, the nations of occupied Europe failed to withhold, eliminate, or even significantly reduce their economic surpluses. The special case of Denmark, a 'model protectorate' until 1943, suggests that a lower degree of coercion was reflected in lower contributions to the German war economy. Thus, although the increasing severity of German exploitation exacerbated hardship and resentment and provoked more resistance of various kinds, Nazi repression succeeded in exacting far greater value in economic resources from Western Europe than the expense of maintaining control. Between 1940 and 1944 European contributions in kind increased by more than 50 per cent the resources available to Germany for the purchase of armaments. 'No slave-driver could have asked for more.'[24]

In sharp contrast, the areas conquered by the Nazis in Eastern Europe, the Soviet Union, and the Balkans were less industrialized and also—in the Soviet case—subjected to deliberate devastation to thwart the invader. Nevertheless their yields of raw materials and food could have been far greater had they not been subjected to the racially motivated policies of plunder, terror, mass-expulsions, enslavement, and genocide. Indeed, some of the subject peoples of the Soviet Union might well have become effective allies of Nazi Germany. The Soviet Union, in particular, had been infinitely more valuable to Germany as a co-operative partner than as a victim. Consequently, Hitler's war of annihilation in the East was 'not only inconceivably evil; it was a strategic blunder which may well have cost Germany the war'. To sum up: the Nazi concept of victory was brutal, unconditional dominion of the most ruthless kind which was almost certain to be self-defeating.[25]

Under increasing pressure from militarist factions, Japan's policy of territorial conquest in Manchuria from 1931 onwards, extended in all-out war to the heart of China six years later, could be broadly viewed as an 'instrument of policy' in Clausewitzian terms. With the advent of war in Europe, Japanese army and naval factions were respectively increasingly tempted to make war against the Soviet Union or the European and American possessions in the Pacific. Hitler's attack on the Soviet Union brought an

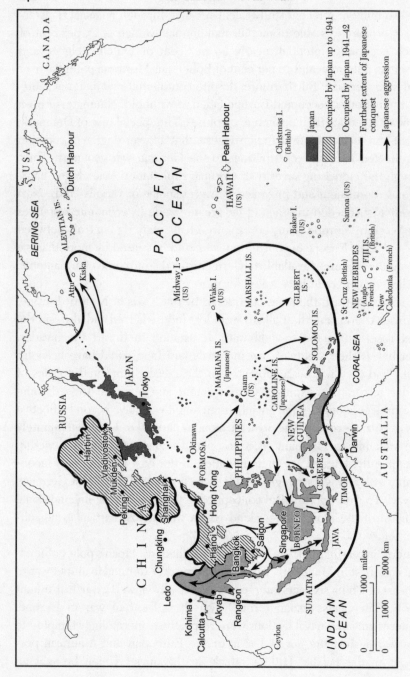

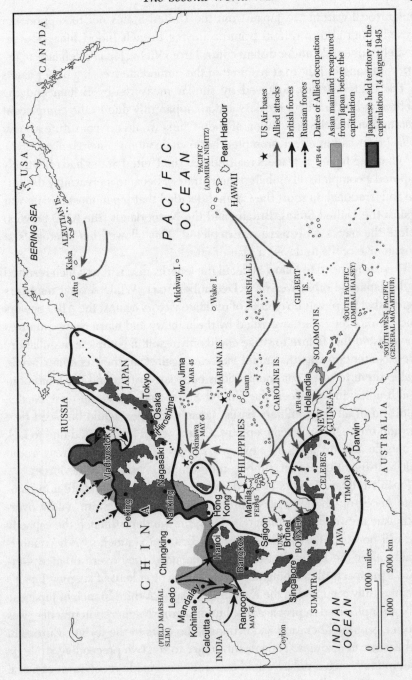

Map 12. Japanese expansion (top) and defeat (bottom), 1941–1945

unequivocal warning to Japan from the United States not to support her Axis ally, but also to refrain from occupying French Indo-China, now extremely vulnerable under distant control from Vichy. Japan's defiance of the latter warning in July 1941 resulted in the immediate freezing of her assets in the United States, followed by similar measures by Britain and the Netherlands. Thus by a reckless action, apparently due to the complacent assumption that the occupation of Indo-China would not constitute a *casus belli*, Japan had in effect prompted her own economic encirclement.[26]

When the Japanese leaders realized that the United States had effectively imposed a complete oil embargo their strategy became increasingly desperate and irrational: in short the Cabinet decided that Japan must initiate war against the United States, Britain, and the Netherlands (the ABD powers) before the navy's oil reserve was depleted. This allowed her strategists at most four months to devise a plan of attack.[27]

There is no need to follow in detail the lengthy discussions which resulted in the Japanese offensives on 7 December 1941. While several ministers argued that there was no chance of ultimate victory against the ABD powers (and China), they were overruled by the military and naval representatives who insisted that failure to strike could only result in complete humiliation. Their contention was that rapid victories against all their enemies would enable them to sever communications between Britain and her Commonwealth allies and between the United States and the south-west Pacific. Deprived of all their military bases, Japan's enemies would be faced by a vast and impregnable Pacific empire. China would be defeated due to lack of outside support and, after two years or so, the conquered areas would yield vast quantities of oil, rubber, tungsten, rice, and other resources.

Japan's military leaders knew they had to deal an immediate and complete knock-out blow or face ultimate disaster. They were wildly overoptimistic in two vital assumptions. First, they underestimated the capacity and will-power of the United States and Britain to launch speedy counterattacks after the initial disasters; and, secondly, they overestimated Germany's prospects of crushing the Soviet Union or at least of keeping Japan's enemies fully occupied in the European theatre. A third branch of Japanese policy, emphasized in propaganda, but disastrously ignored in practice, was to offer genuine liberation and economic benefits to the former European and American colonies. Though subsidiary to the two preceeding strategic assumptions, such a policy would certainly have made reconquest of a real Japanese 'Co-prosperity sphere' much more difficult.

The fatal decision for war made on 5 November thus revealed an appalling failure to think through the likely course of events beyond the assumption of brilliant initial victories. Prime Minister Tojo and his colleagues decided that within two years Japan would be reduced to impotence through the exhaustion of its oil reserves and inability to strike out to seize more supplies. Any hope of holding the territorial gains in China would also have gone. Consequently, despite the enormous risks, the nation must go to war for its survival. Whatever the outcome national honour would be preserved. It was in this fatalistic spirit that an all-out offensive which, by accident, preceded a declaration of war, was launched against the world's most formidable industrial and naval power.[28]

The Japanese forces brilliantly fulfilled their planners' expectations by winning a series of sensational victories at sea and on land. The sinking of the *Prince of Wales* and *Repulse* destroyed British assumptions about the defence of Malaya and Singapore which were both lost in ignominious circumstances by mid-February 1942. Rangoon was occupied on 7 March and the Japanese advanced to the Burma frontier with India, closing the Burma road to China in the process. Wavell's improvised ABDA Command (including American, British, Dutch and Australian forces) quickly collapsed leaving the Dutch East Indies open to invasion. Meanwhile, the extremely daring combined air-sea attack on Pearl Harbor dealt a humiliating blow to the United States, but failed in two crucial respects: it left the oil storage and harbour installations largely intact; and it failed to catch the carriers *Yorktown* and *Hornet* which were absent on exercise. As early as May 1942 the Japanese run of hitherto unbroken successes was checked in the battle of the Coral Sea—a naval encounter yet conducted entirely by aircraft—while in the following month they suffered a critical defeat in the battle of Midway, losing four aircraft carriers to the Americans' one.[29]

The Japanese had won a series of victories not only remarkable in terms of speed and extent but also in economy of force and small-scale losses. The Germans, for example, had advised their ally that it would take one and a half years to capture Singapore and require five and a half divisions: the Japanese accomplished this astonishing feat of arms in only fifty-five days, employing just two divisions. Moreover, so far from needing an overall superiority of numbers against the British, Australian, Indian, and indigenous defenders, the Japanese in the Malayan campaign were outnumbered by approximately two to one. Their total casualties for the capture of

Malaya and Singapore have been calculated as approximately 3,500 killed and 6,000 wounded. According to their own claims, not a single Japanese soldier was taken prisoner.[30] These early victories undoubtedly fostered the national cult of *bushido*, encouraging faith in heroism and invincibility which in the long run would cause both the Japanese servicemen and civilians unnecessary suffering and casualties.

Behind the unconvincing façade of the 'Co-prosperity sphere' the Japanese from the very outset exploited their vast Pacific empire with callous selfishness comparable to the Nazi treatment of occupied Europe. Consequently the vast majority of Asians who experienced their rule found the Japanese army even more oppressive than the European imperialists. Whether, had the Japanese been able to consolidate their empire, they would in time have developed a more humane and coherent programme and ideology for the Greater East Asia Co-prosperity sphere seems doubtful given their racial arrogance. In the event, faced with a deepening strategic crisis, they were content to exploit the conquered economies ruthlessly, while giving only token support to potential anti-European collaborators such as Subhas Chandra Bose.[31]

As with Nazi-occupied Europe, the Japanese military exercised real control through varying forms of national government; ranging from an executive commission of prominent politicians in the Philippines, through the sultanates retaining nominal authority in Malaya, to the puppet administration under Ba Maw in Burma. The economically crucial Dutch East Indies remained under direct Japanese military rule and, as a protectorate, Indo-China enjoyed hardly any more autonomy. Thus, despite some gestures towards a common purpose and a co-ordinated economy, such as a conference held in Tokyo towards the end of 1943, the nationalist aspirations of the conquered territories nowhere approached fulfilment. All that the Japanese succeeded in accomplishing in the long run after their own defeat was the negative victory of rendering the reimposition of European rule so unpopular as to be beyond the military capacity and will-power of the former imperial nations. Japan's failure to win the 'hearts and minds' of any of its conquered peoples was perhaps even more deplorable than Germany's in Eastern Europe since the former rode the crest of the wave of anti-European nationalism (encouraged throughout the Pacific war by the United States), and was dealing almost entirely with pre-industrial and potentially assimilable societies.[32] However, in terms of racial arrogance, brutal suppression of opposition, and callous exploitation of local labour, raw materials, and food supplies, it would be invidious to differentiate between the occupation policies of Germany and Japan.

Could the European war which began in September 1939 have been terminated by a negotiated peace, so obviating the need for a protracted war of attrition culminating in unconditional surrender and the complete victory of the more powerful alliance? In principle, the prospects might appear promising, though it must be remembered that in the later 1930s Britain and France had counted on a long war since that was the only kind they stood a chance of winning. However, in 1939–40 neither Germany nor her opponents had anywhere near mobilized their manpower and resources to full capacity; indeed, Germany actually demobilized substantial forces in the summer of 1940. Nor, though the campaigns in Poland and the West were conducted with great intensity, widespread destruction, and displacement of populations, could they be said to constitute 'total war' such as would occur on the Eastern Front from June 1941 onwards. So, on the basis that neither side was as yet in practice committed to all-out war, it might be supposed that a compromise peace was a real possibility.

Such hopes were ruled out in practice by the incompatibility between Hitler's character, methods, and war aims, on the one hand, and British stubbornness on the other. Paradoxically, the prospects for peace might have been better had Germany's victories in 1939 and 1940 been less complete, because they deluded Hitler and his service leaders into believing that almost any strategic problem could be overcome. Hitler, in Chamberlain's unfortunate phrase, had not 'missed the bus' before launching operations against Norway, the Low Countries, and France, but he *did* miss it during the weeks of May and and June when victory on the continent was clearly being won in a devastating manner. At this critical juncture he gave no serious thought about how to deal with Britain once the defeat of France had left her isolated.[33] Should he offer her an olive branch by recognizing the security of the home islands and the Empire in return for her acceptance of German dominance in Europe; or should he issue immediate orders for invasion plans to be made? Precious weeks elapsed while he dithered, apparently expecting Britain to make the first move towards opening negotiations, thereby acknowledging herself to be the supplicant.

Even under Chamberlain's premiership such an approach was extremely unlikely, both because Hitler was regarded as utterly untrustworthy, and also because no British government in 1940 could have further appeased Hitler by recognizing his territorial conquests in Eastern and Western Europe. Even so, a more serious-sounding peace initiative from Germany, which purported to give Britain a chance to withdraw from the war 'with honour', would have deepened the political division in government circles and, at the least, would have made Churchill's task in rallying the nation in a desperate situation much more difficult. That such a rift existed, though concealed from the public at the time, is evident in Lord Halifax's opposition to Churchill's belligerent stance during the War Cabinet debates on 'A Certain Eventuality' at the end of May, concerning the impending collapse of French resistance, and subsequent secret peace overtures mostly conducted in Sweden.[34] By his indecisiveness during the battle of France and the British retreat from Dunkirk, followed by the belated launching of the battle of Britain in conditions which gave the defenders a good chance of holding out until the autumn, Hitler effectively lost what has aptly been called his 'duel' with Churchill.

Churchill's advent to the premiership on the very day that Germany invaded France and the Low Countries signalled first and foremost Parliament's acceptance of a courageous, belligerent leader who would rally the nation for a supreme defensive effort, and also try to carry the war to the enemy. But who exactly was 'the enemy'? Churchill's premiership soon made plain the harsh reality that is still unpalatable to some students of the war: however loathsome the doctrines and practices of the Nazis might have been, they were not the main reason why Britain had gone to war in 1939. Rather, it was Germany's flagrant infringement of national frontiers, in a seemingly unlimited expansion programme, under a leader who had repeatedly broken his word. In short, Britain was at war with the German state and hence with the Germans, not just the Nazis.[35] The failure to make this distinction was perfectly acceptable to the British people at the time but it would later cause problems in dealing with anti-Nazi Germans and still influences discussions of British responses to the German resistance.

Hitler did not give sufficient thought to his relations with Britain in the summer of 1940 because he assumed she would either come to recognize her untenable position or be defeated in the long run. As early as July 1940 he contemplated invading the Soviet Union in the autumn, but was persuaded by his military planners to delay operation 'Barbarossa' until 1941.[36] At the time this decision seemed less of a reckless gamble than it subsequently appeared. Not only the German generals (with few exceptions) expected a rapid and total victory, but for expert analysts in the West the question initially was for how many weeks the Soviet Union might survive. Between June and October the *Wehrmacht* made deep advances, captured vast 'bags' of prisoners and destroyed enough artillery, tanks, and aircraft as to have ensured the defeat of any other European power. Yet by December 1941 when Hitler had failed to take Moscow and the Russians launched a vigorous counter-offensive, the chances of winning the war in the East had virtually disappeared.[37] Hitler compounded his strategic problems at that time by declaring war on the United States in loyalty to his Japanese ally,[38] thus in effect ensuring that the former would give priority to the European theatre since Germany, as the more powerful enemy, needed to be defeated first.

Much more documentary evidence is likely to be revealed about abortive Russo-German peace feelers when Soviet archives have been more fully opened and explored. It has been suggested that during his loss of nerve in the autumn of 1941 Stalin might have been willing to negotiate, but at that time Hitler was confident of victory. However, from 1943 onwards, many of Hitler's principal subordinates were thinking of trying to end a war which could not be won, some favouring a negotiated settlement in the East, others in the West. As regards the former, Ribbentrop (as well as the Japanese), vainly sought to arrange a truce or peace on the Eastern Front in 1943 and 1944. During the winter of 1943/4 German emissaries took soundings in Stockholm but their efforts were frustrated by their master's unwillingness to admit the hopelessness of the military position.[39]

On the Soviet side, their ambassador in Sweden, Alexander Kollontai, began negotiations with a member of Ribbentrop's staff, and in June 1943 let it be known that they were prepared to discuss a separate peace at a senior level. The Soviet emissaries did not hesitate to reveal to their German counterparts their extreme distrust of their Western allies, who had delayed opening a second front, and were deliberately urging them to fight to their last drop of blood so that an exhausted Soviet Union could be pushed aside at the end of the war. In September 1943 the Russians made a definite peace offer: the Soviet Union would accept the guaranteed frontiers of 1914 and a comprehensive trade agreement with the Reich. Hitler put an end to these negotiations partly for fear of appearing weak, but also because, rightly or wrongly, he thought the Soviet move was really a manœuvre designed to win political concessions from the United States and Britain. In their turn the Western allies knew about these talks but were confident that the Soviet Union had so much more to gain from pressing home her military advantage that she would not accept a compromise peace, or indeed any settlement which left the Nazis in power.[40]

While it may seem surprising that any Russo-German peace negotiations whatever should have occurred, given the ideological thrust behind 'Barbarossa' from the outset and the uniquely barbarous war that ensued, it seems reasonable to conclude that these tentative contacts had virtually no chance of success even if they were genuinely motivated. Neither Hitler nor Stalin was willing to trust the other, for obvious reasons. Stalin had no real incentive to contemplate peace terms once the Red Army began its inexorable drive towards Berlin; while for his part Hitler increasingly took refuge in fantasy, relying on will-power to overcome lack of men and resources, and refusing to acknowledge retreats. Indeed, it has been suggested that Hitler, realizing that he had lost the war in strategic terms, made the 'final solution' of the Jewish problem his overriding war aim from 1942 onwards, mistakenly boasting to his private circle near the end that he had won the 'real' war by cleansing Central Europe of its Jews.[41]

Hitler's increasingly deranged behaviour from 1942 onwards illustrates in its starkest form the ultimate vacuity of the pursuit of victory as an end in itself, when totally divorced from normal political and humanitarian considerations. In view of the evil policies which he had inspired in the East and the consequent mutual war of savagery that had ensued, it is scarcely conceivable that Hitler (or his immediate henchmen) would have negotiated a peace settlement with Stalin. But in terms of national policy, state, survival, and the welfare of the German people, an alternative non-Nazi government needed to end the war in the East by the beginning of 1944 at the latest. Hitler knew this (or should have done if not insane) but remained true to the nihilistic views he had expressed at Nuremberg in 1939: 'we shall not capitulate—no never! We may be destroyed, but if we are, we shall drag a world with us—a world in flames.'[42] This is a world-view which had completely abandoned Clausewitz's ideas of the relationship between war and policy, and by denying the possibility of defeat, and therefore of negotiated terms, invited the total destruction of his own land and people.

In the Pacific, Japan's decision to launch all-out offensive operations against the United States and the European imperial powers left no room for diplomacy—at least in the short term. Moreover, the Japanese leaders' militarist outlook made it virtually impossible for them to give up ground voluntarily or to treat with the Europeans whom they held in contempt. On the other side, what was viewed as the outrage of Pearl Harbor brought America into the war with powerful motives for revenge. Subsequent revelations of Japanese treatment of the wounded, prisoners of war, and civilians generally, as well as their fanatical style of fighting and refusal to surrender, exacerbated fears and hatred to a degree which is now difficult to comprehend.[43] Moreover, in the post-imperial 1990s we must not underestimate the fervent commitment of the dispossessed imperial powers in the 1940s to recover their Pacific possessions even where, as in the cases of the Philippines and India, future independence had already been promised.

The possibility of peace negotiations to end the Pacific war arose very late indeed. During the horrific battle for Okinawa between April and June 1945 the Japanese fought almost to the last man: only a handful of prisoners were captured and more than 100,000 Japanese died. The senior commanders committed ritual suicide. Since the Japanese home islands were now being pounded from the air and they could only retaliate with kamikaze missions, the Americans reasonably believed that significant sections of the enemy high command, as well as senior politicians, would be glad to make peace. But how could the latter open negotiations without being assassinated? No mutinies or even indiscipline occurred in the armed forces and a number of senior officers were determined to fight to the end. The military party, led by the Minister of War, General Anami, and the chiefs of staff of the army and navy, believed that the best chances for a favourable peace would be after the successful repulse of an attempt to invade Japan. Consequently, repeated American broadcasts, signalling that they would respond favourably to any bid for surrender, received no reply from any individual or group with authority to negotiate.[44] The fanaticism of a faction of intransigent militarists prevailed over a widespread public desire for peace and for an end to the city-by-city obliteration carried out by American aircraft from March 1945 onwards.

The Japanese foreign ministry put excessive faith in peace feelers via Moscow, on the false assumption that the Soviet Union would honour its neutrality pact with Japan, not knowing that at the Yalta Conference Stalin had agreed to enter the war against Japan—and indeed was determined to do so. While the Soviets continued to stall, the Western allies issued the

'Potsdam Declaration' on 26 July, warning the Japanese that they were doomed. They must either surrender unconditionally or face 'prompt and utter destruction'. When this final warning was rejected by the military members of the Supreme Council, Truman gave the order to drop the first atomic bomb on Japan.[45]

The Allied determination to win a complete and unequivocal victory over the Axis Powers was encapsulated in the policy of 'unconditional surrender', publicly announced by President Roosevelt at the Casablanca Conference in January 1943 but only making explicit then what had in fact been Allied policy ever since America entered the war. It is hard to think of any other aspect of policy in recent history which has been so widely misunderstood and misrepresented.[46] Contrary to what is so frequently asserted, the unconditional surrender policy had several points in its favour and in reality, as distinct from theory, it did not present Goebbels with a strong propaganda weapon.

The fundamental false assumption of critics of the policy is that, but for 'unconditional surrender', more moderate and acceptable terms could have been offered to the German and Japanese leaders to end the war. But none of the Allied leaders, Churchill, Stalin and, perhaps surprisingly, least of all Roosevelt, ever contemplated anything but very harsh terms towards Germany. These included the partition of the country into several or numerous states, long-term occupation, large-scale indemnities, the trial (or perhaps summary execution) of Nazi leaders, complete disarmament, and the extirpation of the Nazi ideology. In short, the Allied war aims, had they been spelt out, would have been, if anything, more severe than the declared policy and not such as to have encouraged any government capable of further resistance to open negotiations.

Thus, the first advantage of the declaration was that it obviated the necessity to define and state the war aims of the allies, a process which would have been extremely difficult and highly divisive. Secondly, it provided some assuarance to the Soviets that the British and American governments were resolved on the complete defeat of Germany, and would not consider doing a deal with a non-Nazi government (as they had with Admiral Darlan and were to do, despite the Declaration, with Marshal Badoglio). Given Stalin's profound suspicions about his capitalist allies this was a crucial consideration in keeping the Grand Alliance together until both Germany and Japan had been defeated. Thirdly, and the other side of the preceding point, the formula represented an obstacle to the very real German hopes of dividing the allies by concluding a compromise peace either in the East or the West.

While Goebbels certainly welcomed the declaration of the 'unconditional surrender' policy as a gift to Nazi propaganda, he seems in fact to have made very few specific references to it in his subsequent campaigns, as distinct from frequent evocations of the harsh consequences which would follow an Allied victory. Goebbels's unusual restraint in exploiting such a 'gift' apparently stemmed in part from a (justified) apprehension that specific references to the severe terms of the formula would cause the Allies to modify or explain them—as indeed they did on several occasions. For example, Churchill declared in the House of Commons on 22 February 1944:

Unconditional Surrender means that the victors have a free hand. It does not mean that they are entitled to behave in a barbarous manner nor that they wish to blot out Germany from among the nations of Europe. If we are bound, we are bound by our consciences to civilisation. We are not bound to the Germans as a result of a bargain struck.[47]

But common sense suggests that Goebbels's main reason for not exploiting the formula was that he, and the German people, realized that, irrespective of their official formula for ending the war, the Americans, British, and other Western forces would be infinitely preferable as conquerors to the Red Army. Whatever policy towards Germany Stalin might publicly announce, the German people knew very well from 1943 onwards what terrible fate was approaching them from the East.[48]

The most persistent criticism of the unconditional surrender formula is that it caused the Western allies to turn a deaf ear to the German anti-Nazi resisters and thereby prolonged the war. This was an understandable grievance on the part of the plotters themselves, many of whom were individuals of integrity and even nobility of character, but it misses the fundamental point about the nature of the war. This is, to repeat, that the Allies were at war with the German state. The tragedy of the anti-Nazis was that they were also German: 'the anti-Nazi Germans wanted to save Germany, the allies to vanquish it.'[49]

In more practical terms the anti-Nazi plotters were caught up in a vicious circle. They wished to obtain reassurance from the Allies about the way Germany would be treated after a *coup*, since they did not want to be branded as traitors for making possible a 'Carthaginian' peace. But after some disappointing contacts with supposed 'resisters' in the early months of the war, the British government in particular was unwilling to give any advanced commitments before the plotters had demonstrated their ability to act. This, the central dilemma throughout the war, was encapsulated in the tough advice offered by a Chinese revolutionary to Adam von Trott in Berne in 1942: 'If you can't kill Hitler, then kill Göring. If you can't kill Göring, kill Ribbentrop. If you can't kill Ribbentrop, kill any general in the street.' To which Trott replied 'Germans don't kill their leaders.'[50]

We must conclude that the basic objectives of the anti-Nazi resisters were incompatible with Allied war aims. The former wished to preserve a strong German state which would not suffer dismemberment, disarmament, and loss of territory. It would, moreover, retain the frontiers of 1938 (perhaps even of 1939). But the formula of 'unconditional surrender' embodied Anglo-American determination to destroy 'German militarism' (in Churchill's erroneous belief localized in 'Prussia'), and the only group who could have overthrown Nazism were themselves 'militarists' who could not be expected to assist in their own downfall. The alternative to 'unconditional surrender' was a compromise peace with a post-Nazi German government dependent on the army. Even had such an arrangement been made feasible by a successful anti-Nazi *coup*, the arguments against it would still have been formidable. It would almost certainly have entailed an open break with the Russians who were determined to establish their dominance in Eastern Europe. This would have been profoundly unpopular in Britain and the United States, where the image of a friendly Russia under 'Uncle Joe' had been inculcated. It would also have defied the 'lesson' of the failed peace terms of 1919 where the victors had settled for a façade of democracy behind which the unreformed army still exercised real power. Given all that was known and believed about the evil nature of Nazism and its grasp on the German people, it was unthinkable that the Western allies should allow a self-proclaimed democratic government to punish Nazi crimes and bring home to the people that 'the whole nation has been engaged in a lawless conspiracy against the decencies of modern civilisation'. This great effort at ideological re-education could only be attempted through total occupation of the country and full assumption of governmental power. The formula of 'unconditional surrender' was a necessary precondition for such a programme. The victors would need to determine at leisure who were 'good

Germans' fit to govern their country. Finally, it seems clear that the severe implications of the 'unconditional surrender' formula, namely that the Allies would have virtually total power over a defeated Germany, represented not just the personal views of Roosevelt and Churchill (or their concern to preserve the alliance with the Soviet Union), but also the trend of public opinion in their countries, where great efforts and sacrifices had been demanded in the pursuit of victory.[51] The full horror of the post-war revelations of the Final Solution programme would serve to confirm popular views in the West that Germany must be treated severely for the Nazis' evil regime.

Discussion of the Allied pursuit of victory in the Second World War requires a brief reference to problems of grand strategy. In the Soviet case, the shock of sudden invasion by a nominal ally, and the ensuing German advances in 1941 and 1942, left Stalin very little choice. First he must win the defensive struggle for sheer survival, accompanied by relentless demands for Allied support, direct and indirect. He must then wage ceaseless attrition, symbolized by the siege of Stalingrad, and must conclude by a ruthless ground and air offensive to carry the war deep into the Balkans, East Europe, and the Reich itself. Soviet naval strategy was of secondary importance and the air force did not develop an independent strategic bombing offensive of any significance. Since the Soviet Union was not at war with Japan until August 1945 it was saved from the problems of a two-front war.

By contrast Anglo-American strategic decision-makers faced dilemmas and differences of opinion which have provided grist to the mills of historical controversy ever since. In principle, though some Americans (most notably Admiral King) did their best to thwart it in practice, Roosevelt and Churchill agreed to concentrate on defeating Germany first and never wavered on this decision. Both historical tradition and the shattering experiences of 1940–1 caused the British to opt for a gradual, constricting strategy, referred to as 'tightening the ring' round Germany, by a combination of economic blockade, strategic bombing, and 'setting Europe ablaze' by encouraging anti-Nazi resistance movements in occupied countries. It was optimistically hoped that these strategies, greatly assisted by anti-Nazi opposition on the home front, might so weaken German military power in the long term that the eventual second front in Western Europe would be a mere 'mopping-up' operation. Unfortunately for Britain, none of the three branches of this strategy had any significant effects in the short term. Britain—and from November 1942 the United States also—were distracted from the main task by an agreement to defeat Italy in the Mediterranean; and Stalin, abetted by influential supporters in the West, made repeated demands for an immediate opening of a second front in Western Europe. Perhaps most important of all, American strategists were never entirely convinced by British insistence on a peripheral strategy or indirect approach. Their military tradition, doctrine, rapidly growing supremacy in weapons and material, and not least their anxiety to finish off Germany quickly so as to concentrate on the war against Japan, all pointed to a rapid and direct confrontation with the German forces in Western Europe as the prelude to a drive towards Berlin.

There is no need here to rehearse the resulting Anglo-American strategic controversy in any detail.[52] Whether wisely or not, Churchill and the chiefs of staff prevailed until the launching of *Overlord* on 6 June 1944 in their cautious and pessimistic views that the cross-Channel invasion must not be attempted until it was certain to succeed; that Allied military and French civilian casualties must be kept to a minimum; and that, man for man at the tactical level, German soldiers were superior to any of the Allies.

This British reluctance to take risks in a continental campaign explains to a considerable extend the great faith which the British government and its air advisers—assisted by the Americans from early 1942—placed in the strategic bombing of Germany as a means not just to soften up the enemy (and wreak vengeance) but to win the war. This strategy remains one of the most controversial aspects of the Allied war effort. Historians who have studied the issues agree that the campaign fell a long way short of winning the war on its own, but they disagree on many other aspects, such as whether the main bombing offensive could more profitably have been diverted to other targets (as it was, successfully, in preparing the way for *Overlord* in 1944); whether more precise targeting would have brought better results; and, overall, the significance of its contribution to the eventual winning of the war against Germany.

An intrinsically complicated issue is often further confused by a failure to keep moral objections and strategic criticisms strictly separate. A few individuals consistently opposed all strategic bombing on moral grounds, but many others who have entered this controversial area seek to condemn the programme when it was demonstrably ineffectual and indiscriminate, while accepting the principle of bombing when it was 'precise' and effective. Some critics leave the impression that they would like the war to have been won without hurting the Germans, but that was not a widely held view at the time. In reply one may argue that it was only by persevering with a strategy known to be ineffective that better techniques for finding and hitting targets could be developed; that the cumulative effect was significant, though hampered by adverse weather, enemy defences, and changes of priority in targets; and that 'precision bombing' was and still remains a relative term. There is no consensus, even among specialists in the subject, on the signifi-cance of the Allied strategic bombing campaign's contribution to the victory over Germany. This account endorses the positive, though necessarily quali-fied conclusions reached by Ian Kershaw and Richard Overy.[53]

The policy of strategic bombing was bitterly criticized in the early years of the Second World War because its advocates had claimed far more for it than it could fulfil. After Dunkirk and the battle of Britain it provided the only means by which Germany could be directly attacked, but the practical results were extremely disappointing—as was reluctantly admitted in 1941. Not until 1943 was Britain prepared in technical and organizational terms to mount a serious and sustained offensive. Thereafter, in terms of the Allies' general strategic intentions: 'Bombing achieved all that was expected of it. Only those who expected bombing to win the war on its own were frustrated by events.'

In so far as the strategic bombing of German cities was intended to break German morale, to the extent of causing a collapse of the domestic war effort, it manifestly failed. But it would be a mistake to assume that in cities or areas which were intensely bombed, such as the Ruhr, Hamburg, and Berlin, civilian morale was not seriously affected. On the contrary, bombing was the most important factor in causing a decline in morale. The Nazi leadership failed to shift increasing popular disaffection, growing into hatred, from themselves to the enemy. What had not been thought through was the fact that, under a totalitarian system, the population was strictly controlled and had no opportunity whatever to convert war-weariness and even defeatism into political action. Draconian measures were necessary to suppress public criticisms of the regime and even of Hitler himself.[54]

Of even greater significance, however, was the effect of bombing on the German economy. The indiscriminate 'area' attacks of Bomber Command, although they interrupted the smooth running of the economy and harassed the workforce, failed to destroy significant capital stock that was difficult to replace. But by 1944 the Allied air attacks *were* beginning to have significant effects on such crucial aspects of the economy as the oil industry, steel production, and aircraft manufacture.

Perhaps of even more importance in the long term were the indirect effects which bombing caused by diverting resources from other military roles. By 1944, for example, some two million soldiers and civilians were engaged in the ground anti-aircraft defences—more than were employed in the whole aircraft industry. In the same year Speer estimated that 30 per cent of total gun output and 20 per cent of heavy ammunition was intended for anti-aircraft defences. In principle Hitler opposed the use of the *Luftwaffe* in home defence but in the summer of 1944, under pressure from Speer, he agreed to divert 2,000 fighter aircraft to protect vital fuel installations and industries.[55]

Wartime intelligence, apparently confirmed by the post-war Strategic Bombing Surveys, indicated that the German war economy, so far from declining under the impact of strategic bombing, was in fact expanding at a faster rate than before until 1944. But, as Overy points out, bombing was 'much more effective than the Allies believed', in that it prevented the increase from being considerably greater than it was. In 1944, for example, Hitler set a target of 80,000 aircraft a year but in fact only 36,000 were produced, and even this shortfall does not indicate the full story since many new factories were destroyed and there were heavy losses in assembled aircraft, fuel, and spare parts. Above all we should contrast the disruptive effects of the bombing on Germany industrial planning and production with the complete freedom enjoyed in the United States to 'plan, build and operate the war economy without interruption and as near to the economic optimum as possible'. In conclusion, the Allied strategic bombing offensive began to pay dividends in terms of industrial destruction (especially of fuel supplies) and disruption of transport just at the point when Germany was poised to produce large increases in war material beyond those already planned. Consequently, but for the culmination of the bombing offensive in 1944 and 1945, Germany would have been far better equipped to prolong and intensify the final battles for her homeland, perhaps thereby raising the possibility of the use of the atomic weapons in Europe, for which they had originally been intended.[56]

Of all the controversies centred on the strategic bombing campaign in

Europe, that concerned with Dresden has been the most persistent and intense. Indeed, the destruction of the Saxon city, famous for its architecture and porcelain, is frequently used as a moral exemplar on which to condemn the whole bombing strategy. The reasons for this fixation are obvious. Dresden was a beautiful city, previously subjected to only small-scale American attacks in October 1944 and January 1945; the war in Europe was nearly over (though this was not so apparent at the time either in Germany or in southern England where the V2 rockets were raining down); the destruction and civilian deaths were unexpectedly severe; and the raids appeared to have no clear strategic purpose. What added fuel to the flames of the Dresden controversy, however, was Churchill's very belated revulsion against an operation which he had strongly urged on the air staff, hence Portal's successful insistence that his critical minute be withdrawn and modified.[57]

Tragedies and disasters are inseparable from the conduct of war and Dresden was only significant for the reasons mentioned above. Seventy German cities had been attacked by Bomber Command and many of them suffered greater destruction than Dresden, notably Hambury, Dusseldorf, Berlin, and Cologne.[58]

Whether they are now deemed to be adequate or not, there were definite strategic reasons for the attack on Dresden early in 1945. As early as August 1944 British political authorities and the air staff discussed a suggestion that an overwhelming blow on a relatively undamaged city might have such a devastating effect on civilian morale that it would force the high command to end the war. The original target selected was Berlin, but in February 1945 the choice fell on Dresden. By this time, however, there was little hope that one such devastating air attack on a city would end the war.[59]

By the end of January 1945 the allies were considering night area bombing of east German cities which were also communication centres, in connection with a desire to assist the Russian advance into Germany. Though not specifically requested by the Russians, Dresden was one of the cities included in the Allied list. On 26 January the Prime Minister prodded the air staff to attack east German cities and the Secretary of State for Air duly informed him that Dresden was among the targets selected. Thus, as the Official History comments, neither the air staff nor Harris can justly be accused of waging war in a different moral sense from that approved by the Government. Webster and Frankland go further by remarking that, although the area offensive is open to many strategic criticisms, 'it is difficult to see why it should bear unfavourable moral comparison with naval blockade or some other kinds of warfare'.[60]

On the night of 13 February just over 800 aircraft of Bomber Command attacked Dresden in one of the most devastating raids of the war. American Eighth Air Force carried out three further attacks on the city. Some of the survivors of Dresden and their descendants still understandably express bitter feelings towards Bomber Command (though not it seems the American Eighth Air Force), but do they also feel shame and remorse towards their compatriots who were daily committing unspeakable deeds in the concentration camps? Nor should the bomber crews who attacked Dresden on orders stemming from the highest political authority feel regret or shame for this particular raid on a target as legitimate as many other cities which had previously been devastated.[61] It was regrettable that, through a combination of circumstances, British bombers inflicted excessive casualties on enemy civilians, but it is a quirk of polemical historiography to focus obsessive attention on Dresden at the expense of Hamburg, Cologne, and Berlin, not to widen the discussion to cities flattened earlier by the *Luftwaffe*. The real tragedy of Dresden was that it took place long after a more rational and humane German leader would have ended hostilities, to prevent precisely this kind of slaughter and destruction in a war that Germany had clearly lost but the Allies had not yet conclusively won.

'Hiroshima' (though not, curiously, Nagasaki) has become such an emotionally charged word, symbolizing the horrors of war and a propaganda centre as a 'mecca for world peace', for anti-Americans and extremists of both right and left in Japan,[62] that it is very difficult to write dispassionately about how a definite Allied victory was clinched in the Pacific war. What is too often forgotten, most conspicuously by the Japanese themselves, is that the dropping of the two bombs (the only ones available) came as the *coup de grâce* of a very long, widespread, and notoriously brutal war in which, fanned by propaganda, national and—it must be stressed—racial hatreds had been whipped up. It is also too easily forgotten, particularly in light of the following decades in which 'deterrence' was the key concept, that atomic weapons were developed in a race against the Germans *to be used*, originally in Europe but, as events turned out, in the Far East.

There is no need here to explore in detail all the issues raised by Truman's decision to employ atomic weapons to end the war against Japan.[63] The main controversies are familiar to all students of the war. Would the Japanese leaders have achieved an effective surrender had they been given more time? Should the retention of the Emperor have been more emphatically excluded from the 'unconditional surrender' policy? Should the cities selected as targets have been given prior warning? Why was it necessary to drop a second bomb? Would the invasion of the Japanese mainland have been as difficult and costly as American estimates assumed? Was Japan only the incidental victim with the real but covert aim to deter the Soviet advance into Manchuria and give America ascendancy in the opening phase of the cold war?

Truman was well aware of Japanese peace feelers via Moscow, but at Potsdam Stalin confirmed that the enemy's emissaries were unwilling to accept the Allied terms. The President also knew that his military advisers were anxious to avoid the invasion of Japan. Plans were being made to land in Honshu in November 1945 and near Tokyo in March 1946. The defenders were known to have more than two million well-armed troops available in strongly fortified positions. It was unlikely that their resistance would be less stubborn and fanatical than at Iwo Jima and on Okinawa. The war might well last another year and casualties would be numbered in millions. Given this forecast and the fact that Japanese cities, particularly Tokyo, were already being annihilated by fire storms, Truman's decision was not a difficult one. None of the key American policy-makers questioned the use of the bomb, there was almost universal support for it on the Allied side and, in public at any rate, Truman never expressed doubts about his decision.[64]

On 6 August a 14 kiloton uranium bomb was dropped on Hiroshima, killing some 90,000 people and destroying much of the city. The Emperor urged the government to end the war regardless of conditions, but the army refused and tried to suppress all information about the effects of this new weapon. On 8 August Soviet Union declared war on Japan and the next day her forces attacked Manchuria. This prompted the Emperor to summon the Supreme Council and again urge that the Allied terms be accepted, but the War Minister and chiefs of staff opposed, insisting that Japan must not be occupied, that the armed forces should be allowed to return home and be demoblized under Japanese supervision, and that any suspected war criminals should be tried by Japanese courts. This military opposition was not changed by the dropping of a 20 kiloton plutonium bomb on Nagasaki on 9 August. When, a few days later, the Emperor finally took the initiative by announcing that it was his will that the war be ended, officers from the War Ministry and general staff attempted to isolate the palace and seize the recording of the peace appeal. They were overpowered and the leaders committed suicide in the traditional manner. On 15 August the Emperor's message was broadcast and on 2 September the formal instrument of surrender was signed aboard the American warship *Missouri*.[65]

Although the Soviet Union's declaration of war undoubtedly strengthened the hand of the Emperor and the Japanese political leaders who favoured peace, it is hard to dispute the common-sense judgement that Japan had been brought to the brink of defeat through conventional bombing, to which the atomic bombs provided the dramatic and horrific finale. Whereas the case of Germany remains disputed, the fate of Japan shows that, in suitable conditions, strategic bombing could deal the knock-out blow to end the war on its own—that is, without the involvement of ground forces.

Speculation as to whether and when conventional bombing by itself would have forced Japan to surrender and whether an invasion would have been necessary may be fascinating for those who wish to criticize Truman's decision, but he had to act in the light of intelligence available at the time, including predictions about prolonged enemy resistance. The militarists' attempt to invade the imperial palace and prevent the Emperor's peace broadcast provides an insight into what might have occurred, in Japan itself and in all the occupied territories, but for the cataclysmic effects of the atomic bombs. Churchill recorded in his memoirs that he immediately realized that the apparition of this 'almost supernormal weapon' might provide a pretext for the Japanese people to save their honour and so avoid being killed to the last fighting man'.[66] And, as a Dutch prisoner of the Japanese in Batavia, faced with almost certain execution, Laurens van der Post confirmed that the dropping of the first bomb had precisely this effect. In a prisoner-of-war camp in Nagasaki an RAF doctor, Aidan MacCarthy, and his colleagues had actually dug their own graves and were saved from certain execution by the dropping of the second bomb which either killed their guards or caused them to flee in terror.[67] Throughout Japanese-held territory, including Burma, where the invasion of Malaya was being planned, hundreds of thousands of soldiers rejoiced at the sudden ending of the war. Countless Asian civilians and Japanese soldiers also owed their lives to the sudden termination of the war.

The case of Japan in 1945, perhaps even more so than Germany, represents an extreme departure in modern times from the concept that war is an instrument of policy which is waged in the interests of the state and its inhabitants and must be terminated when those interests are clearly not being served. As a historian sympathetic to Japan has written, the 'spirit of the nation had passed into the custody of the patriotic societies' which would have murdered anyone who dared to speak of surrender. It had become virtually impossible for Japan to capitulate. 'Japan, having made a cult of the principle that no Japanese ever surrenderd to the enemy, now found it impossible to accept the findings of common sense.'[68]

In the light of the enormous superiority in industrial power enjoyed by the United States, the Soviet Union, and their principal allies, including Britain and Canada, over Germany, Japan, and their allies it may seem surprising that it took so long to achieve victory. On the contrary, it could be argued that, since several interrelated conflicts were taking place more or less simultaneously on a global scale, and since the aggressors had achieved control of vast populations and resources by their conquests in 1939–42, it was a remarkable achievement by the victors in terms of organization, war production, and strategic planning to conclude both the European and Pacific wars in 1945. Indeed, it has been suggested above that, without the dropping of the atomic bombs, the Pacific war might have lasted into 1946 and even beyond. The senior historian of British intelligence in the war has argued that the Allies' access to massive and absolutely reliable high-level intelligence (i.e. 'Ultra') shortened the European war by at least three years.[69]

Even with the avoidance of hindsight, it may be suggested that it was clear to the Axis from about mid-1943 onwards that they could not win the war in the ways that had seemed possible at the peak of their victorious expansion. Yet only Italy, faced by invasion, opted to change her leader, cease fighting, and change sides, incidentally avoiding the full imposition of the unconditional surrender terms. In contrast, the governments of Germany and Japan, both authoritarian and militarist, but differing greatly in organization and ethos—and the latter having no dominant individual comparable to Hitler—alike refused to think seriously about ending the war until their cities were being destroyed and their people suffering almost beyond endurance. Their opponents certainly employed ruthless methods, not only in strategic bombing, and their motivation doubtless included hatred and the desire for revenge. War, however, is a ruthless activity and in the 'total' conditions of 1943–5 it is hard to see why the stronger powers should have been more moderate towards the rampant aggressors of 1939– 42 whose abominable policies and practices were becoming widely known. The only way the European and Pacific wars could end was by the latters' acceptance of the terms of unconditional surrender.

It is commonplace to remark that long, general wars often fail to resolve the initial issues for which they were started, while raising others that bedevil the winners, even to the extent of making victory itself seem futile. The outcome of the Second World War notoriously provides rich pickings for the cynics and ironists. Thus, for example, Britain emerged on the winning side but exhausted herself in the process and had soon to relinquish most of her empire and with it her great-power status. France, a loser in 1940 but a member of the victorious alliance in 1945, suffered more intensely and longer than Britain due to her experience of defeat and occupation, and also lost her empire after a prolonged and painful struggle. The Soviet Union was an undisputed winner on the battlefield, her status as one of only two world powers was confirmed, and she acquired through direct or indirect control a vast glacis or buffer zone in Central and Eastern Europe. But the enormous losses and physical destruction she had endured left her exhausted and relieved rather than rejoicing: it was certainly not the kind of success expected from the resort to war as an instrument of policy, rather a matter of survival. By contrast, the United States was an unequivocal winner: her industry and commerce and 'cultural imperialism' enormously boosted by war; her homeland completely secure and undamaged; and her military power overwhelming in Western Europe and Asia. If the military and economic burdens of world power were soon to prove heavier than anticipated, they could nevertheless be borne with some comfort.

Further ironies soon became evident in the post-war fates of the two principal losers. Japan, under American occupation and tutelage, and West Germany under occupation and with considerable economic stimulus from Marshall Aid and additional Allied assistance, rapidly became more prosperous than any of the victors except the United States. But the greatest irony, perceived by acute observers to be impending from 1943 onwards,[70] was that the West (and above all the United States) devoted enormous efforts to defeating Germany and extirpating Nazi tyranny only to create a vacuum in Central and Eastern Europe which was filled by a comparable tyranny with an even more threatening, because more insidious, ideology.

However, the substitution of Soviet for Nazi dominance in the centre of Europe, accompanied by forty years of cold war, does not mean that the West fought on the wrong side or that the outcome was futile. Liddell Hart now has few supporters among serious historians in arguing that victory was 'an illusion',[71] which is very different from saying that victory failed to produce all the results desired by those who had fought for ideals such as democracy, self-determination, and a more secure world order. A. J. P. Taylor, in general no admirer of military power or of war as an instrument of policy, concluded his illustrated survey of the Second World War with the following remarkable sentences: 'Those who experienced it know that it was a war justified in its aims and successful in accomplishing them. Despite all the killing and destruction that accompanied it, the Second World War was a good war.'[72] This judgement may to some people savour of hindsight and parochialism, more convincing to a citizen in London, New York, or Bonn, than in Warsaw, Budapest, or Sarajevo. Also, some would question whether it had been a 'good war' throughout. It had begun largely in terms of *realpolitik* to stop German, Italian, and Japanese territorial expansion, but had increasingly acquired idealistic and ideological overtones due to the United States' dominant role and revelations of the sheer evil of the Nazi and Japanese regimes, epitomized by the nightmare images from Belsen, when liberated in April 1945. In retrospect, therefore, it was a great and noble achievement to overthrow regimes which constituted 'the worst challenge ever presented to liberal civilization and its conception of the humane society'.[73]

Moreover, though the post-war rooting out and punishment of former Nazis in West Germany was far from perfect, the Allies' valiant efforts at re-education and rehabilitation may generally be accounted successful. West Germany, with the exception of small aberrant minorities, has until the tensions recently caused by reunification, been a model democracy which has firmly rejected Nazi-style bullying and militarism. Under her own sobering experience, reinforced by post-war American guidance, Japan too appears to have learnt beneficial lessons from her defeat. Without the near-total control afforded to the democratic victors by the terms of unconditional surrender, can it be asserted with any confidence that West Germany and Japan would have followed the prosperous and peaceful courses that have characterized their history since 1945?

It is therefore cynical, careless, or plain foolish to say that wars settle nothing and particularly that victory in the Second World War was an illusion. Had the Axis Powers not been unequivocally defeated and their authoritarian-militarist regimes destroyed, the post-1945 history of Europe and Asia would have been vastly different—and darker—than it was. To anyone growing up in the West after that watershed the Allied victory probably seems inevitable, but that was not how the prospects looked— despite Churchill's brave rhetoric—before the German failure in Russia and the Japanese attack on Pearl Harbor. Whatever the shortcomings and disappointments of the eventual victory, particularly Anglo-American impotence in the face of the Red Army's occupation of Eastern Europe, cynically exploited by Stalin, they were infinitely preferable to defeat. In this negative sense war had also been an instrument of policy on the part of those, mainly democratic, states who had been forced into hostilities in self-defence and in the interests of establishing a more secure world in which they could flourish.

No war achieves a perfect solution even for the victors, and settlements imposed and sustained by armed force do not last forever—witness the remarkable collapse of the Soviet Union and its satellite empire in Eastern Europe and the Balkans. In 1945 it was sufficient to celebrate VE and VJ days in the knowledge that Europe had been saved from Nazism and Asia from Japanese imperialism. That the world the celebrants inherited was 'divided, grimly austere, profoundly unjust and threatened by nuclear destruction does not mean that they were misguided'. Western Europe and the free world generally had gained a reprieve—a reprieve which they are still enjoying half a century later.[74]

The Pursuit of Victory in the Nuclear Age

John Keegan concludes his stimulating study of *The Face of Battle* with the reflection that 'the suspicion grows that battle has already abolished itself'.[1] What he had chiefly in mind was the great projected clash of the super-powers and their rival alliances in Central Europe in which nuclear weapons would probably be used. Writing in the mid-1970s, Keegan believed that modern conditions of war, generally, let alone of battle itself, had become too psychologically stressful to be endured by the majority of troops. Confidence that battle was meaningful and decisive was waning; future battles might well be fought only in 'never never land'.

Whether or not the sheer horror and unendurability of late twentieth-century combat conditions provide a full and satisfactory explanation, Keegan would surely have been right about the unlikelihood of large-scale conventional wars if he had referred specifically to Western Europe, Scandinavia, and North America. Unfortunately, the existence of nuclear weapons, and massively destructive conventional armaments, have by no means ruled out wars and large-scale armed conflicts in many parts of the world. A standard reference work covering the period 1945–89 lists one hundred such conflicts, many of them still continuing.[2]

Until the astonishing collapse of the Soviet Union and its Communist client states in the late 1980s, governments, analysts, and commentators were understandably preoccupied with the dangers of nuclear war and how to deter it, though many military men and strategists continued to believe in the possibility of large-scale conventional operations. For the superpowers, however, the risks of provoking a nuclear exchange were widely held to outweigh any possible gains, while for their allies and satellites there was the obvious danger of being drawn into a local conflict. Consequently, a 'balance of prudence' was created, in which conditions most strategists accepted that the only sure 'firebreak' was the dividing line between conventional and nuclear weapons.[3] Such deductions were not, of course, simply the products of theory. The Korean War, the Cuba missile crisis, and the dangers of supporting rival belligerents in the Middle East—all impressed on the superpowers the risks of war by proxy. Hence there developed a mutual recognition of 'no go' areas where conventional wars must be quickly stifled by superpower pressure if not prevented outright.

It is doubtful, however, if the existence of nuclear weapons and their role in superpower rivalry alone accounts for the decline in the incidence of large-scale conventional wars between states in some parts of the world where they had previously occurred quite frequently. Even without the additional risk posed by nuclear weapons, it seems clear that the price paid for accepting the security afforded by alliance with a superpower is a readiness to subordinate national interests to the needs of the superpower relationship. This was a lesson which Britain and France learnt painfully in the Suez crisis. Similarly, China received no support from the Soviet Union in its attempts to seize Quemoy and the Matsus from Taiwan in 1955 and 1958.[4] Furthermore, although there are innumerable territorial disputes throughout the world, the United Nations Charter expressly forbids the use of force to alter national boundaries. This prohibition may provoke a cynical smile in a decade which is witnessing the drastic reordering of the former Yugoslavia and many parts of the former Soviet Union by force, but in other regions it has proved extremely difficult to have conquests recognized by the international community.[5] Some states, such as Israel, have been content, until very recently, to inhabit the juridical twilight zone between war and peace, but it should not be assumed that such defiance has no international disadvantages, quite apart from generating domestic tensions.

One may also generalize that in democratic states, and particularly those with conscription, it is not easy to persuade the electorate to support a conventional war, short of a direct attack, and extremely difficult to retain that support for a prolonged war in which heavy casualties are incurred. It must be doubtful if Britain would have attempted to recover the Falkland Islands had her forces had been dependent on conscription. Moreover, in an era in which popular attitudes are heavily influenced by television, governments in free and open societies are put under enormous pressure to intervene to prevent massacres, starvation, and anarchy, but severely criticized when the lives of soldiers or of 'innocent civilians' are forfeited as a consequence.

Another consideration is that, despite all their imperfections, friction, and competition, international trade and financial interdependence makes warfare too uncertain, if not actually counter-productive, for developed, industrial states. This should have been evident to Germany's rulers before they embarked on two world wars, but is generally recognized now—and more to the point—embodied in international institutions and agreements. If in some respects they were too idealistic, optimistic, and polemical, Sir Norman Angell and other pre-1914 proponents of the futility of war on economic and financial grounds are now seen to have been broadly right—or at least to be pointing in the right direction.[6]

Above and beyond all specific arguments against the resort to war in pursuance of national interests, it may be suggested that among developed industrial states—mostly but not invariably democracies—it is simply assumed that offensive war is unacceptable for a congeries of legal, moral, and practical reasons. How else can one explain the euphemistic transformation of Ministries of War into 'Ministries of Defence' and the blanket employment of the word 'security', even when referring to areas characterized by mayhem and anarchy? If all armed forces are 'defensive' where is the threat? Clausewitz would have been unimpressed by our obfuscating addiction to euphemisms.

Perhaps, however, it is Clausewitz, with his passionate commitment to the state, cool acceptance of war as an instrument of policy, and preference for victory through concentration of forces for the decisive battle, who is outmoded? Each age rediscovers and remoulds its version of the Prussian theorist to meet contemporary needs; his outstanding work *On War* being sufficiently dense, complicated, and suggestive as to permit widely varying interpretations and extrapolations.[7] While historians have stressed the depth but also the limitations of his personal experience and insights into the political and military conditions of his own times, strategic analysts (on both sides of the Iron Curtain) have enthusiastically propounded his relevance for the nuclear era. Three aspects of his theory of war have seemed to them particularly worthy of updating. First, there is his teaching on the relationship between war and policy; namely that 'War is only a branch of political activity; it is in no sense autonomous . . . it cannot be divorced from political life'.[8] Moreover, policy must permeate and *control* strategy, the latter having no independent existence of its own. This may seem a truism but the political context was often lost to view in the alarming scenarios of the high priests of nuclear theology. Secondly, without the need for much manipulation, Clausewitz's ideas could be usefully applied to the complex doctrine of deterrence. He had, after all, in his early chapter on 'Purpose and Means in War', noted that strategy could have—and indeed frequently did have—the negative object of making clear to the other side 'the improbability of victory . . . (and) its unacceptable cost'. He also realized the importance of 'unfought engagements', which is also an aspect of nuclear deterrence.[9] Thirdly and perhaps most important, were Clausewitz's insights into why, in practice, most wars are likely to be *limited*. They would be limited by the cultural and social constraints of the age, by the goals of policy, and by war's inherent tendency to friction, inertia, or tedious inactivity. As Michael Howard points out, Clausewitz did not live to develop his notion of the 'two types of war', limited and unlimited, which could more realistically be said to cover the whole spectrum of organized violence. But he remained acutely conscious from his own military experience that it took the consent of both parties to keep a conflict 'limited'.[10] No matter that his ideas on this crucial issue were not fully worked out. They appealed much more to nuclear-age theorists than his alleged advocacy of 'total war', which had been fiercely criticized after the First World War.

The traditional belief in the possibility of 'victory' in warfare has been most seriously undermined by the existence of nuclear weapons in control of more than one power. Until 1953, when Eisenhower effectively quashed the notion, it was possible, though never really convincing, to believe that the West possessed a war-winning instrument, or rather one which might coerce the enemy into submission without war. The meaning of 'victory' in a war between nuclear powers was hard to imagine from the outset, and became even harder as both sides developed sophisticated delivery systems and defences with a view to achieving mutual deterrence. What was perhaps less predictable, but rapidly became accepted in practice after the United States refrained from using nuclear weapons in the Korean War, was that such overdestructive weapons were unlikely to be used against a non-nuclear opponent either, except perhaps in extreme circumstances where the nuclear power's very existence was at stake. Even these circumstances have been difficult to describe convincingly but this element of uncertainty is held to be crucial to the effectiveness of deterrence.

Whereas in the West it was unfashionable to 'think the unthinkable' by discussing the prospects of victory in a war with the Soviet Union, Soviet textbooks and strategists consistently refused to distinguish between a deterrent and a war-fighting capability. Nuclear forces were included in the supposedly Clausewitzian imperative that military forces 'have no rationale save as instruments of state policy', with the corollary that Marxist-Leninism would prevail over the capitalist enemy whatever weapons-systems or policies he might adopt. Soviet reliance on a first-strike capability against American land-based inter-continental ballistic missiles (ICBMs) and their well-publicized civil defence programme, though in retrospect seen to be exaggerated by a few Western strategists, gave some credence to the alarming view that the Soviet Union expected to win even if nuclear weapons were used.[11]

Whether this was all part of the game of bluff and counterbluff in the interests of deterrence, monotonously repeated due to the constraints of Marxist-Leninist orthodoxy, or whether—despite their unique losses and destruction during the Second World War—Soviet leaders really did believe in military victory over the West, was fortunately never put to the test. But around 1980 it provoked Colin Gray and others to argue that Bernard Brodie and the orthodox American proponents of deterrence were, in effect, defeatist. The West too should be fortfied with a theory of military victory. As recently as 1990 Colin Gray has asserted that 'Provided that a state has an absolute end in view—the definitive elimination of a Carthage, a Corinth, a Jewish revolt, a Nazi Germany, a Soviet Union—absolute means are not incompatible with strategy'. By 1961–2, he argues, the United States enjoyed a truly splendid first-strike capability. 'Although Western Europe probably would have suffered severely, there can be little doubt that a U.S. (and British) surprise attack on the U.S.S.R. . . . would have achieved a truly definitive military victory.'[12] There is no need here to deploy the counter-arguments, ably developed in an earlier controversy by Michael Howard. The fact remains that responsible Western leaders did not believe that a meaningful military victory over the Soviet Union could be achieved. Instead, with a rapidity and completeness that surprised virtually all Western experts, the Soviet Union collapsed through the breakdown in its economy which had for many years been overstrained and was eventually broken by the intensive arms race with the West.

The continuing relevance of the concept of victory in conventional wars is more problematical both because the motives of states resorting to war are disputable and also because the cases are too diverse to indicate a clear trend. As was noted earlier, deliberately preparing for war as 'an instrument of policy' has acquired additional hazards in the nuclear era, but still it does occur. North Korea's invasion of the South, the Anglo-French attack on Egypt in 1956, Argentina's seizure of the Falkland Islands, and Iraq's occupation of Kuwait all seem to be fair examples. As for the states invaded or otherwise subjected to aggression, it is both reasonable and legitimate that they should strive for victory, if only for the limited aim of expelling the invader or recovering lost possessions.

The best prospects for the state contemplating military aggression in late twentieth-century conditions would seem to be to launch a pre-emptive strike for clearly defined, limited objectives, with a view to wrapping up operations in a few hours or, at the most, days. Where the adversaries are fairly well-matched in such criteria as divisions, armaments, and morale then a clear-cut victory is likely to remain as elusive as at any time during the whole period surveyed. But the devastating use of superior air power, as demonstrated by Israel in 1967 or the United States in the Gulf War in 1991, shows that a knock-out blow in blitzkrieg style is still possible in theory. Otherwise, a clear-cut decision in lengthy conflicts is likely to be frustrated by exhaustion of manpower, arms, and equipment, decline in popular support, and, most likely of all, outside interference. Stalemate, as in Korea or the Iran–Iraq conflict, is a more likely outcome than a clear admission of defeat. If a United Nations mandate for 'unconditional surrender' could not be adopted against Saddam Hussein's Iraq it is unlikely to be enforced anywhere.

A protracted conflict which *did* end in a definite result was that in Vietnam, due to the United States' limited stake in the outcome and the revulsion of its own public opinion. Even so, the United States sought to escape from Vietnam in 1973 with an honourable draw comparable to the outcome in Korea, but two years later, in 1975, the completeness of its humiliating defeat was rubbed in by the unification of the country under a merciless Communist regime.

It has taken an avowed admirer of Clausewitz, Martin van Creveld, to produce a truly radical critique of the relevance of the master's theories in the late twentieth century. He is unhappy, fundamentally, with Clausewitz's assumption that organized violence should only be called 'war' if waged by the state, for the state, and against other states. As he points out, the modern European-dominated states system dates only from the mid-seventeenth century (the Treaty of Westphalia in 1648 to be precise), and is by no means universal, comprehensive, or necessarily enduring. In fact, he sees the state's ability to monopolize military power and maintain order as faltering and even failing under the threat of terrorism and other forms of violent protest. Should present trends continue, the Clausewitzian model of war based on the division between people, army, and government seems, in his opinion, to be on its way out. Unless low-intensity conflict can be quickly contained, 'it will end up destroying the state'.[13]

Van Creveld makes a strong case that, since 1945, the states system with its traditional military organization has proved increasingly unable to deal with the wide spectrum of violence, terrorism, and armed conflict below the level of conventional war. The preoccupation with nuclear deterrence, which has greatly inhibited the willingness to risk inter-state wars, has resulted in the establishment of overly destructive and excessively costly 'dinosaurs' which are totally unsuitable to deal with low-intensity conflict. While there is much truth in this as a generalization about the cold war era, particularly as regards Central and Western Europe, recent events warn against the dangers of projecting past trends into the future. As a reviewer of Van Creveld's English edition pointed out, the collapse of the Soviet Union seems to have made the world safe for the use of conventional weapons.[14] The Gulf War demonstrated that there are other adversaries, armed (mostly by the West) with sophisticated modern weapons, including long-range rockets, against whom conventional warfare can be waged without risk of nuclear escalation.

Let us outline Van Creveld's other main criticisms before considering qualifications and counter-arguments. As mentioned above, Van Creveld sees low-intensity warfare as the main and growing threat to the orthodox states system and military organization. As an Israeli citizen he has doubtless been disturbed by his government's inability to suppress the *intifada*, though to be fair he discusses other cases where regular forces have failed, including the Americans in Vietnam and the Soviet Union in Afghanistan.[15] He rightly stresses that in such protracted struggles there is virtually no place for the Clausewitzian decisive battle; indeed, the exception of Dien Bien Phu suggests that such a set-piece encounter plays into the hands of the revolutionary enemy, since the insurgents can afford to wait until confident of victory.

Van Creveld subjects Clausewitz's central notion that 'war is a continuation of politics with an admixture of other means' to a rigorous analysis. He

points out that, historically, wars have been fought for other reasons than
'policy'; that the state's 'interests' are not necessarily those of the people
who have to do the fighting; and that the rational, calculating notion of war
as an 'instrument of policy' does not fit the facts of many modern conflicts—
particularly those where national survival is at stake.

Finally, he queries the role of 'the people' in Clausewitz's trinity. He
believes that in most of the wars ever waged the vast majority of combatants
lacked any precise idea of the political considerations for which they were
supposed to be fighting. On the other hand, he waxes lyrical, even ecstatic,
about the supreme and unique appeal of war as a kind of intensified sport
in which the individual faces the ultimate challenge in the risk of being
killed as well as killing.[16]

Clausewitz's bleak world of competing nation states has an almost nostal-
gic appeal compared with Van Creveld's nightmare vision of an imminent
future in which such organizations as the IRA, PLO, ETA, and Hizbollah
will call the tune, kept in check only by gendarmerie, militia, and police with
dubious legal and moral status.[17] While Van Creveld correctly notes the
decline in the number of conventional inter-state wars, there are still
enough to be significant to justify the retention of regular forces, albeit
better equipped and trained to fight limited wars as distinct from a compo-
nent in the deterrence of nuclear war. Making predictions in print is a
hazardous business and Van Creveld was unfortunate to include in the
American edition of his book, published on the eve of the Gulf War, the
reflection that, 'all things considered the Iran–Iraq War may well have been
among the last the world will see' (p. 18). Of course this may still prove
correct in the long term, but in the mean time the Gulf War has given his
critics plenty of ammunition. Even so, it was probably typical of things to
come in being waged between an independent state and a United Nations
coalition, in contrast to the Falklands conflict which saw the clash of two
sovereign states from different continents.

On the critical issue of low-intensity warfare, Van Creveld does not
sufficiently distinguish between revolutionary movements able to secure
mass support and aspiring to seize political power, as they did successfully,
for example, in Cuba and Vietnam; and smaller-scale movements or groups
perpetrating acts of sabotage and terrorism but unlikely to be more than an
irritating nuisance to stable, resolute, and efficiently defended states. One
might go further and suggest that the tide of revolutionary warfare may

already be ebbing. Writing a decade ago, Michael Howard argued that its effectiveness had been limited to the fragile structures of post-colonial and similar societies, 'and it has yet to be shown that it can really be an effective technique in international conflict between major powers'. More recently, in a brilliant essay, John Shy has reinforced Howard's scepticism by stressing the terrible costs as well as the poor prospects of revolutionary warfare waged against any but the most feeble regime. 'In China and Vietnam, revolutionary war meant millions dead and a generation of suffering for millions more; the brutal discipline required for revolutionary endurance stretches the powers of comprehension.'[18]

The state system is also proving more resilient and popular than its critics allow. True in Western Europe some governments, though not necessarily their electorates, seem anxious to surrender their sovereignty, but throughout the former Soviet Union and the countries it occupied or dominated, such as the Baltic states, there is a marked eagerness to acquire even the trappings of independence. Furthermore, whether it is a matter for regret or rejoicing, sovereign states do, at least in principle, retain the right to pursue war as an instrument of policy. Contrary to Van Creveld's opinion, it may also be contended that war in self-defence for national survival is still an act of policy. Czechoslovakia exercised its (terminal) right as an independent state in deciding not to oppose Germany in 1938, whereas Poland opted for resistance the following summer. Clausewitz covered this dilemma in his sardonic remark 'The aggressor is always peace-loving . . . he would prefer to take over our country unopposed'.[19] On this and on other critical issues of terminology, Clausewitz certainly requires careful interpretation in the light of greatly changed modern conditions, but his ideas about the state, war, and politics are by no means obsolete.

In particular, students of war in the nuclear era such as Bernard Brodie, André Beaufre, Alastair Buchan, and their successors have extended the scope of strategy beyond the operational aspect to concern themselves with broader questions about the role of different forms of force in international politics. Consequently, the modern content of strategy covers not only war and battles but also 'the application or maintenance of force so that it contributes most effectively to the achievement of political objectives'. Thus 'strategy' is involved in any contest of opposing wills even if not a shot is fired in anger. The French term 'total strategy' perhaps best conveys the wider modern usage which is applicable in peace as well as in war and can

employ pressures other than the purely military, including legal, psychological, moral, and, perhaps most important, economic.[20] Clausewitz was by no
means ignorant of, or indifferent to, some of these additional, often indirect
and more subtle, methods of bending an enemy's will to one's own, but he
certainly gave precedence to the bloody decision by battle. Since he was
scrupulously concerned with historical accuracy, he would surely concede
that decisive victory in battle is not only much harder to achieve in modern
conditions but also in most circumstances entirely inappropriate to the
furtherance of the national interest.

This chapter will conclude with a brief survey of the main conventional
wars since 1945, not for the purposes of narrative or description, but rather
to illustrate through case-studies the problems relating to the pursuit of
victory in an era characterized by the fear of nuclear war, by the disappear-
ance of European dominance, and by the vastly extended membership of
the United Nations—from forty-five states in 1945 to 183 in 1993. In sum,
we are left with only a limited number of areas where the superpowers and
the United Nations are unwilling or unable to intervene; where there are
inter-state disputes which seem worth fighting for; and where there is a
suitable 'war theatre'. While it would be tempting fate to say that 'total
victory' in the sense of the defeat of the enemy forces leading to occupation,
annexation, or the destruction of the state is now impossible, it is safe to say
that it is rarely aimed at and extremely hard to achieve.

Perhaps the main significance of the Korean War (1950–3) in terms of
this study is that it demonstrated that a large-scale but limited war could be
fought to a stalemate in the atomic era. The first American troops (in what
was of course technically a United Nations operation) landed at Pusan
airport on 1 July 1950 and their commander, General MacArthur, fully
expected to have won the war, by driving the North Korean forces back
across the 38th parallel, by Christmas. However, the advance of the UN
forces into North Korea on 1 October led to the entry of China into the war
on a massive scale. Seoul fell to the Chinese on 4 January, only to be
liberated two months later.

By this time MacArthur, supported by Republican sympathizers at home,
was chafing at the 'unparalleled conditions of restraint and handicap' imposed on him, claiming that victory could not be won in Korea unless he was
allowed to wage all-out war against China. In an incautious letter to a
Congressman, MacArthur argued that the Korean conflict was World Com-

munism's attempt to gain power and that, in fighting it, the United States was defending Europe. If the war in the East was lost the fall of Europe would be inevitable. 'We must win', he concluded. 'There is no substitute for victory.'[21] This final gesture of defiance led to his dismissal by President Truman on 10 April 1951.

The Chinese entered the war with the limited aim of driving back the UN forces to a respectable distance from the Yalu river. But they too were deluded by their remarkable advance towards the end of 1950 into the belief that they could obtain a crushing victory. Consequently, they pushed south too rapidly, overextending their supply lines, only to be decisively checked by the reorganized UN forces now commanded by General Matthew Ridgway. Each side having abandoned its original limited aims, it took two further years of desultory fighting (July 1951–July 1953) before the Panmunjon armistice recognized the existing front line as the border between the states of North and South Korea.

The Communist powers had carefully avoided a direct confrontation with the United States; the Chinese forces operating in Korea were officially referred to as volunteers, and Soviet ground forces were not involved. Even the Soviet Union's indirect support was cautious; her supplies only began to reach China in bulk in the autumn of 1951 and even then every ton had to be paid for.[22]

The support of the United Nations for such a large-scale military operation—as distinct from peace-keeping activities—was only made possible by the Soviet boycott of the Security Council and would not be repeated for the remainder of the cold war. The outcome was clearly a diplomatic compromise reflecting the military stalemate: the UN forces had failed to reunite Korea but had frustrated the North from overrunning the South. From the later perspective of failure in Vietnam this was quite a satisfactory achievement by the United States, but at the time the national reaction was rather one of frustration at discovering the political limits of military power in the new international order. MacArthur's belief that the unleashing of America's superior military power was bound to bring victory was widely supported and is not entirely dead now. Many Americans in positions of authority drew the erroneous conclusion that, given sufficient military and economic support, almost any anti-Communist regime could be kept in power, regardless of its own merits or popularity.[23]

India and Pakistan have fought three wars since gaining their independ-

ence in 1947, as well as indulging in a good deal of skirmishing short of
formal hostilities. The first two wars (1947–9 and 1965) were fought essentially over disputed boundaries in Kashmir. Both were brought to a speedy
end by United Nations' diplomatic pressure, and after the second, the
Declaration of Tashkent (10 January 1966) restored the pre-war *status quo*,
leaving the ceasefire line in dispute. Though costly for the local people (the
1965 war cost about 20,000 lives, mostly civilians), this is an exceedingly
protracted and localized dispute which poses no great threat to the international order—unless other powers should intervene.

The third war (in 1971) was more significant in that it involved large-
scale, intensive combat, reached a decisive result, and had international
repercussions. In an attempt to crush the Bangladeshi movement for independence in East Pakistan, after a year of tension and fighting, Pakistan
launched a pre-emptive air strike (on 3 December 1971) in an attempt to
knock out the Indian air force on the ground, but this failed. India re-
sponded with counter air strikes and a five-pronged invasion of East Pakistan totalling about 160,000 troops. The land operations culminated in a
two-day battle for Dhaka before the Pakistani commander surrendered (16
December). Simultaneously to the west, on the border of Jammu and the
Punjab, India successfully fought one of the largest tank battles since the
Second World War (Pakistan losing forty-five Pattons and India fifteen
Centurions), before declaring a unilateral ceasefire.[24] India made no
territorial claims against what up till then had been West Pakistan, but the
latter was obliged to recognize Bangladesh's independence. Pakistan, which
was far more dependent on outside military support than India, received
substantial backing from the United States, particularly after the Soviet
invasion of Afghanistan, in addition to her existing links with China, whereas
India signed a twenty-year Treaty of Peace, Co-operation, and Friendship
with the Soviet Union.

Earlier, following the Chinese occupation of Tibet in 1959, a Sino-Indian
dispute arose over two mountain areas on their new mutual frontiers at
Aksai Chin and Ladakh. In June 1962 Indian forces moved forward in an
attempt to occupy both disputed areas. A swift and effective Chinese riposte
followed in October, involving, in the west, a 30-mile drive into Ladakh and,
in the east a 100-mile advance beyond the McMahon Line. Having demonstrated their superiority by defeating an Indian counter-offensive, the
Chinese declared a unilateral ceasefire, withdrew about 12 miles in each of

the disputed areas and returned all captured Indian equipment. Although the borders remain in dispute, this trial of strength exposed serious weaknesses in the Indian forces and redounded greatly to the credit of the Chinese for their skilful combination of force and diplomacy in a limited war. The same cannot be said for China's ruthless occupation of Tibet and the brutal repression of its people.

The Middle East provides the outstanding example of an area where conventional forces have repeatedly demonstrated their utility in large-scale conflicts. In the campaigns of 1967 and 1973 particularly, the Israeli forces achieved brilliant, 'decisive' victories, at least equal in terms of planning, mobilization, tactics, and strategy to any others discussed in this volume.[25]

The Arab–Israeli conflicts also illustrate the ambiguities of the term 'limited war'. The formal wars have been of short duration and confined in area; the combatants have been restricted to a considerable extent by dependence on foreign powers for the supply of weapons and equipment; and, at least in the Israeli case, by a commendable determination to economize in casualties. The Israelis have throughout held the 'total' war aim of preserving their state and, up to and including 1967, the positive but still limited aim of extending their territory to enhance the prospects of self-defence. By contrast, the participating Arab states have had the unlimited aim of the destruction of the state of Israel, but in every instance to date have had to acknowledge defeat and cut their losses. Egypt has already achieved something approaching normal inter-state relations through the Camp David accords in 1979. She has recognized Israel's right of existence and, apart from the special problem of the Gaza Strip, the pre-1967 frontiers between the two states have been restored. A new phase of peace negotiations between the Israelis and representatives of the Palestinians began in 1993 and is making gradual progress towards substantial measures of autonomy for the Gaza Strip and the Israeli-occupied West Bank.

It cannot be overemphasized, in accounting for the ruthless, rapid, and decisive nature of Israel's campaigns, up to and including 1973, that they have been regarded—in the most literal sense—as wars for survival. This essential fact largely explains the people's willingness to make onerous sacrifices in the interests of defence, the generally excellent state of war readiness of the standing forces and the remarkable system of citizen reservists, the professional skills and high morale of the services, and the government's preparedness to launch pre-emptive strikes. By contrast their

Arab enemies have often fought well, especially in defence, but have lacked determined, flexible leadership and resilience in the face of tactical reverses. A tradition of defeat and failure had become so established that, even at the start of the Yom Kippur War in 1973, when the Egyptian forces in particular achieved surprise and impressive early successes, their aims—originally limited—became overambitious, and they soon lost the initiative. Above all, Israel's opponents have failed to achieve a true alliance and have been unwilling to co-operate closely in policy, strategy, or military organization.

Here we propose to concentrate on the 1967 or 'Six Day War', not only because it brought Israel's most spectacular military victories, but also because it resulted in immense territorial gains which enhanced Israel's security but also created problems which only now, in the mid-1990s, seem to be moving towards a peaceful resolution. On the eve of the Six Day War in June 1967 Israel's predicament looked desperate, if not hopeless, surrounded and outnumbered as she was by the forces of Egypt, Jordan, and Syria with the support of Iraq. Yet within a week, in what has been called 'a miracle of deliverance', she had decisively defeated all her opponents in the conventional military sense.[26] The 'miracle' owed much to an astonishingly successful pre-emptive air strike against the most immediately threatening opponent, Egypt, who had concentrated 80,000 troops and 1,000 tanks in Sinai. Israel's strike forces, comprising 800 tanks and 350 combat aircraft, had been built up with foreign aid, notably from France. In the early morning of 5 June, in 500 sorties which evaded the Egyptians' Russian-supplied radar defences, 309 out of 340 serviceable Egyptian aircraft were destroyed, mostly on the ground, including nearly all her modern Russian bombers and fighters. By the evening of 6 June a large proportion of the Syrian, Jordanian, and Iraqi air forces had also been destroyed. The Egyptian army in Sinai was well dug in in strong defensive positions some 100 miles east of the Suez canal, but without air cover it was doomed. Operating in three mechanized columns in a plan reminiscent of German blitzkrieg, the Israeli tank divisions broke through on the first day, after hard fighting, and by the third day (7 June) had joined up behind the bulk of the Egyptian forces at the Mitla and other passes. This brilliant manœuvre caused a disorganized retreat in which most of the Egyptian vehicles were destroyed or abandoned.

Already the Israelis were aware of desperate international moves to secure a ceasefire and were determined to secure defensible military frontiers before the United Nations intervened. In some of the hardest fighting in any of her wars the Israelis expelled the small but well-trained Jordanian forces from the narrowest point of the country around Jerusalem, and on 7 June achieved their main goal of capturing the Old City and the Temple Mount. Finally, before at last heeding insistent international demands for a ceasefire on the evening of 10 June, the Israelis drove the redoubtable Syrians off the whole plateau of the Golan Heights from whence their northern territory had been under close observation and sporadic shelling.

In relation to its population (less than four million) and strategic vulnerability (hardly 12 miles wide at the narrowest point) Israel had achieved one of the most complete victories over four enemies in the whole history of warfare. The defeated states' war-making capacity had been temporarily destroyed and all had suffered heavy casualties: Syria had lost about 5,000 men killed, Jordan had lost 6,000 (and the strategic buffer zone of the West Bank, including Jerusalem), and Egypt at least 15,000 dead. Israel, by contrast, had lost fewer than 800 troops killed. The immediate rewards of victory were impressive: in tangible terms the Golan Heights, the whole of the West Bank, and the vast Sinai peninsula rendered Israel for the first time a relative secure and defendable state. Intangibly she had won enormous military and political prestige in the eyes of the Western powers and was now regarded as the most valuable ally in the region.

There can be few more vivid and illuminating examples of the ambiguities and ironies of military victory than Israel's experience since her peak of achievement in 1967. If, in some quarters, hubris resulted from her sudden accession of power and glory, nemesis has not yet followed: indeed, in 1973 the Israeli forces again inflicted humiliating defeats on Egypt and Syria, after themselves suffering a surprise attack and initial reverses which brought the nation to the brink of disaster. The stark fact remains, however, that Israel has not been able to translate her military successes into a stable peace settlement with her neighbours—with the important exception of Egypt and, to a lesser extent, Jordan—or with the million additional Arabs in the occupied territories. Instead she has experienced what a sympathizer has aptly called 'the tragedy of victory'.[27] Arthur Hertzberg even suggests it might have been better for Israel had the Six Day War ended in a draw rather than a series of stunning victories. These victories gave Israelis a feeling of euphoria and a sense of power far exceeding anything they had

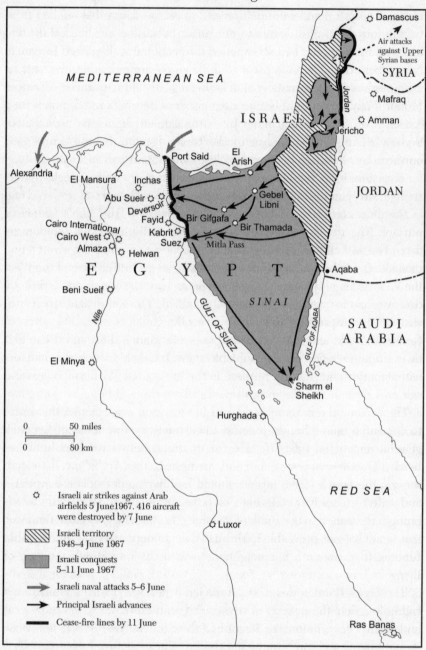

Map 13. Israeli victories and conquests in the Six-Day War, 1967

MEDITERRANEAN SEA

SYRIA

Air attacks against Upper Syrian bases

Damascus

Mafraq

ISRAEL

Amman

Jericho

Jordan

JORDAN

Port Said

El Arish

Alexandria

El Mansura

Inchas

Abu Sueir

Gebel Libni

Deversoir

Fayid

Bir Gifgafa

Bir Thamada

Cairo International

Kabrit

Cairo West

Suez

Mitla Pass

Almaza

Helwan

E   G   Y   P   T

Beni Sueif

SINAI

Aqaba

SAUDI ARABIA

Nile

GULF OF SUEZ

GULF OF AQABA

El Minya

Sharm el Sheikh

Hurghada

RED SEA

0   50 miles

0   80 km

✡   Israeli air strikes against Arab airfields 5 June1967. 416 aircraft were destroyed by 7 June

Israeli territory 1948–4 June 1967

Israeli conquests 5–11 June 1967

Israeli naval attacks 5–6 June

Principal Israeli advances

Cease-fire lines by 11 June

Luxor

Ras Banas

enjoyed before. In the immediate aftermath of victory Eshkol had been willing in principle to negotiate the return of all the occupied territories except Jerusalem, but had received a flat rejection from the Arab League in November 1967. Thereafter fears that Israel's victories could turn out to be even more poisonous than defeat have been partly borne out. The American Jewish lobby for a time became more strident and influential, providing a 'blank cheque' for Israeli religious extremists and aggressive nationalists, thereby making a moderate political settlement more difficult. From 1968 onwards Israeli settlements on the West Bank made the handing over of any territory there to the Palestinians an even more emotive political issue, though one which is now being courageously confronted.

Israelis also reflected bitterly on their treatment at the hands of the great powers. In 1956 Dayan's brilliant conquests in the Sinai had been over-turned at the conference table, while in the Yom Kippur War it was noted that no outside power had intervened so long as the Israelis were losing, but the Americans and Russians both did so to save the Egyptians from total defeat on their own territory after General Sharon had crossed the Canal and cut off their armies. Hertzberg finds a direct link between the victory in June 1967 and the attempt to repeat the success in a 'short and cheap' war by invading Lebanon in June 1982. Whereas Israel's previous wars had met with general sympathy in the West as being essentially defensive, the inva-sion of Lebanon was viewed with similar consensus as being an aggressive attempt to end the Palestinian problem by force. It is not the intention here to apportion blame between Arabs and Israelis for the failure to secure a general negotiated peace settlement. If the Israelis have been guilty of behaving with condescension and arrogance, the Arabs for their part seem sometimes to have forgotten that they have repeatedly attempted—and failed—to destroy the state of Israel by resort to war. Perhaps the point has been reached when Israel needs peace more than the Arab states, always excepting the Palestinians, dependent as she is on American funding to maintain her readiness for war as well as a high standard of living.

The United States was never formally at war with North Vietnam, but rather fought in the capacity of adviser and protector to a succession of weak and corrupt governments in Saigon. This was, of course, no mere technical-ity but a fatal handicap because the United States was never able to establish a regime capable of commanding the positive allegiance of the South

Vietnamese people or of an army which would protect them against terrorism and intimidation. True, the United States (under the United Nations flag) had succeeded in preserving an equally corrupt and anti-democratic regime in South Korea, but there she enjoyed two advantages conspicuously lacking in Vietnam. The Korean peninsula provided no neutral safe-havens for Communist infiltration or regrouping after set-backs; and the North Koreans and Chinese played into American hands by waging what was essentially a conventional war. By contrast, in Vietnam geography and the nature of the terrain favoured guerrilla operations as an instrument of revolutionary war. Moreover, while it would be wrong to assume that a Vietcong victory was inevitable, the revolutionary nationalist movement had outstanding political and military leaders in Ho Chi-Minh and Vo Nguyen Giap respectively, who already had acquired vast experience and confidence through their successful campaign to expel the French.[28] Above all, time was on their side in a struggle which was total for the Viet Minh and the National Liberation Front but distant, limited, and hard to understand for the American electorate.

The original American goal was indeed the admirable one of creating the social, economic, and political conditions in which South Vietnam could enjoy democracy and prosperity, but from the outset her representatives displayed a deplorable lack of understanding and sensitivity in handling the local people. Whether there was a 'right military strategy' which the Americans might have adopted may be doubted because it seems probable that such a strategy—for countering revolutionary warfare—was simply incompatible with American political culture and military traditions. However, it is possible that the air offensive against North Vietnam (operation *Rolling Thunder*) would have succeeded had it been waged all-out from the start in 1965, when the North's defences were flimsy, rather than gradually increased over two years, But this operation presumed a stable South Vietnam government, which was never achieved. What can be asserted with confidence is that Washington fundamentally misread the nature of the struggle from the outset. It entertained a theory of victory—to be won essentially by orthodox land and air operations—which would force the enemy to accept a negotiated settlement favourable to South Vietnam.[29]

The rapidity and scale of the build-up of American troops in Vietnam gives a good indication of the growing desperation in Washington to end the war by victory through overwhelming numerical and material strength. By the end of 1961 there were some 3,000 American personnel in South Vietnam, four years later it had risen to 181,000, and by 1966 to 385,000. In 1968 the Communist 'Tet' offensives against southern cities were defeated, but so unpopular was the war that President Johnson decided not to seek re-election and to begin peace negotiations. In January 1969 American troops in Vietnam reached a peak of 541,500, and thereafter there was an equally rapid withdrawal until the last personnel left in March 1973. Some 55,000 American soldiers and 5,000 allies died in the war. Total Vietnamese deaths exceeded one million.[30]

It would be mistaken to generalize from the Vietnam War, as so many anti-American or pro-Communist theorists did in the 1970s, that revolutionary insurgents were bound to triumph over orthodox military forces. The British success in Malaya and the Americans' own performance in the Philippines showed that this was not necessarily so. But in Vietnam, in addition to the self-imposed handicaps mentioned earlier, the United States was paradoxically disadvantaged by its excess of fire-power, bombing aircraft, and intelligence; that is, there was less incentive for a patient build-up of political reform and military confidence in the South. Such expedients as the bombing of the northern cities and the 'body count' suggest that the Americans were not able to counter the enemy's skills and sheer ruthlessness in the use of intimidation, terror, and propaganda. A less publicized but probably more significant factor was the very broad American consensus existing in 1965 that the Communist threat in South East Asia must be stopped, which suddenly evaporated due to the Communists' failure in Indonesia and China's absorption with the Cultural Revolution. In cold war terms, 'Vietnam ceased being clearly vital to American interests within a year after the decisions to send troops were made.'[31]

Although the prevailing American academic view now seems to be that its own and other Western reporting of the war, in newspapers, on radio and, above all, on television, was not primarily responsible for undermining public support and eventually bringing about a humiliating withdrawal, the powerful indictment by veteran Far Eastern reporter, Robert Elegant, is not easily set aside. In Elegant's view the media became the main battlefield: illusory events reported by the press as well as real influences at work within the press corps were more decisive than the clash of arms. In his opinion, the outcome of the war was determined not on the battlefield but on the printed page and, above all, on the television screen.

Looking back coolly, I believe it can be said (surprising as it may still sound) that South Vietnamese and American forces actually won the limited military struggle. They virtually crushed the Viet Cong in the South, the 'native' guerrillas who were directed, reinforced, and equipped from Hanoi; and thereafter they threw back the invasion by regular North Vietnamese divisions. None the less, the War was finally lost to the invaders after the US disengagement because the political pressures built up by the media had made it quite impossible for Washington to maintain even the minimal material and moral support that would have enabled the Saigon regime to continue effective resistance.

Elegant attributes much of the 'surrealistic reporting', to several factors including ignorance of Vietnamese language and culture, isolation from the quixotic American army establishment, physical and mental distance from the military conflict, short tours of duty, and political prejudice in favour of the North Vietnamese 'liberators'. He mentions several journalists who later expressed shame at their biased reporting, including a West German, Uwe Siemon-Netto, who wrote, after five years covering the war: 'I am now haunted by the role we journalists played over there.' Why, he asked rhetorically, did we not report the evil nature of the Hanoi regime and its numerous atrocities? And what prompted us to make our readers believe that the Communists, once in power in all of Vietnam, would behave benignly? Elegant's answer is that the majority of reporters and commentators followed the fashionable line that it was their duty to be 'critics of the American war'. His extremely well-documented article concludes bleakly that:

Despite their own numerous and grave faults, the South Vietnamese were, first and last, decisively defeated in Washington, New York, London, and Paris. Those media defeats made inevitable their subsequent defeat on the battlefield. Indo-China was not perhaps the first major conflict to be won by psychological warfare. But it was probably the first to be lost by psychological warfare conducted at such great physical distance from the actual fields of battle—and so far from the peoples whose fate was determined by the outcome of the conflict.[32]

In the mid-1970s America's failure seemed complete and her demoralization profound. In contrast to Korea, the army's reputation plummeted. As a veteran of both wars remarked: 'We went into Korea with a very poor army, and came out with a pretty good one. We went into Vietnam with a pretty good army, and came out with a terrible one.'[33] Not until the Gulf War did the American services recover a good deal of their self-confidence and public reputation, both so seriously tarnished in Vietnam.

Yet in the longer term the meanings of victory and defeat in Vietnam are suffused with irony. The American belief in the domino theory proved erroneous: Communist triumph in Vietnam did not reverberate through South East Asia and Africa. The victorious regime has been isolated, impoverished, and repressive, condemned for its violations of human rights by both Amnesty International and Asia Watch.[34] Thousands of so-called 'boat people' have risked extremes of suffering, death, and imprisonment to escape the delights of 'the people's democracy'. The highest ambition of many who remain in Hanoi and other northern cities is to experience freedom and prosperity like the Japanese or, one might add, the South Koreans and Taiwanese. One British journalist who reported the Vietnam War has recently made a return visit to the country and has had the courage to admit that he and his colleagues presented a distorted picture by vilifying the Americans and idealizing the Vietcong. He concluded with a speculation unthinkable even a decade ago: might it not have been better if the Americans had won?[35]

The Falklands War seemed anomalous when it was fought in 1982, in that the Foreign Office had given clear indications that British sovereignty over the islands was negotiable, and a decade or so later looks even odder, as an old-fashioned inter-state conflict over a very distant and rather impoverished colony—though this could be dramatically changed if reports of oil finds '50 per cent bigger' than the North Sea field prove correct. But for the military historian it presents many interesting aspects and its outcome, in terms of the pursuit of victory, was decidedly positive.

The Argentine Junta evidently resorted to war as a calculated instrument of policy on the assumption that Britain would not attempt to reverse a *fait accompli*, and that success would greatly increase its own popularity. In operational respects it was a limited war, in that combat exclusion zones were proclaimed and that Britain refrained from attacking Argentinian ports and air bases. On land both governments desperately needed a clear-cut victory: the British because of their initial blunder in foreign policy and because they took an enormous gamble in dispatching the Task Force; the Junta because it needed 'the Malvinas' to survive in power. Despite frantic international efforts to mediate there was no room for compromise on the issue of sovereignty.

These issues were decided by conventional amphibious battles, culminating in the land battle for Port Stanley. All the British services emerged with credit, but only the Argentinian air force enhanced its reputation. Britain's resort to war was generally supported in West European countries; she had acted with a clear UN mandate against blatant acts of aggression and with invaluable, albeit covert, support from the United States.

The British victory was so complete that it could not be disputed or muffled, and the results were almost entirely beneficial for both sides—a very rare occurrence in military history. The Junta was deposed and its leading members tried and punished, to be replaced by a more democratic regime. The British services (all volunteers) proved their value to the nation and enjoyed a surge of popularity whose ripples are still in evidence. The government's decisive and resolute leadership throughout the crisis probably helped it to win a further term in office. At the time, though this deduction now looks dubious, it was widely assumed that Britain's successful action heralded a brave new era in which acts of aggression would be punished by 'the international community'. On the debit side, cynics will point out, Argentina has not abandoned her claims to the islands, while victory has saddled Britain with a vast annual expenditure for posessions which she must now retain for some time as a matter of pride. But in contrast to the post-war history of Vietnam, where an American victory would at least have brought greater prosperity to the South, there seems, nothing to be said on behalf of Edward Luttwak's sour and provocative comment that it would have been better for Britain if both her aircraft carriers had been sunk and she had lost the war.[36]

The Iran–Iraq War (1980–8) was, in some respects, even more surprising than the Falklands conflict, in that it was extremely long, attritional, and very reminiscent of the First World War in its mainly static and episodic character. It ended in stalemate and mutual exhaustion. By the end of the war each side had between 1.3 and 1.6 million men under arms. The specific issues which led Iraq to attack Iran were so parochial as to appear almost derisory to all but the belligerents, concerning as they did the precise demarcation of the rival claims to the Shatt al Arab waterway, and control of the Iranian border province of Khuzestan. There were also fundamental religious differences between the two Muslim states and Saddam Hussein feared subversion from radical Iran. Ultimately, however, there was rivalry for the dominant role in the area, particularly as regards control of oilfields and port facilities. Both sides had received a substantial part of their armaments from the West, which helps to explain the intensity of sporadic offensives, but also the lulls and the failure of either side to sustain mobile operations which might have secured a clear victory.

The war began with a full-scale Iraqi invasion in September 1980 which captured Khoramshahr but was held up just short of Abadan, the main Iranian oil centre. An Iranian counter-offensive between May 1981 and June 1982 recaptured nearly all the lost ground. At this point there seemed to be a good opportunity for peace negotiations, but Iran now invaded Iraq. Thereafter Iraq conducted a largely defensive war against annual Iranian spring offensives, using poison gas in the process. Between 1983 and 1988 Iran launched her main offensives in the south but narrowly failed to capture Basra. Both sides bombed the other's oil installations, and hostilities also extended to a tanker war in the Gulf which caused several Western nations to send in warships to protect their interests. Iraq derived financial support from the Gulf States and received arms supplies from the USSR, and from the West, notably from France. In the final year of the war, both sides launched offensives: the Iranians in the north against Iraqi Kurdistan, where they came close to seizing the reservoir supplying Baghdad; while in the south the Iraqis regained the Fao peninsula and Majnoon island. In July 1988 the exhausted belligerents accepted a UN-sponsored ceasefire and withdrew to pre-war boundaries. Relations between the two states remained hostile until, in mid-August 1990, in order to secure his Iranian flank while annexing Kuwait, Saddam Hussein handed over all Iranian territory still in his control and also accepted the terms of the UN Security Council (Resolution 598) which he had previously ignored. Casualty statistics for this fanatically fought war were certainly high but imprecise, estimates varying between a total of 750,000 and one million dead, with Iran's losses much heavier then Iraq's.

It is hard to explain why the war dragged on for so long, particularly after the *status quo* had been restored in 1982. Iran, with her superior resources of wealth and manpower seemed more capable of victory, but Iraq was prepared to make the greater effort by mobilizing a much higher proportion of her manpower and keeping her forces better equipped and trained. Iraq made ruthless demands on other Arab countries for assistance (only Syria and Libya failed to support her), and also won substantial aid from other powers outside the region. Iran, by contrast, displayed a marked inability to

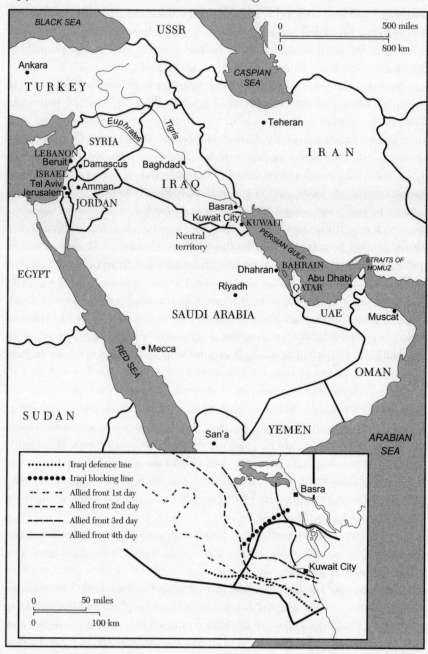

Map 14. The main areas of conflict in the Gulf War

conciliate or make friends. She wantonly alienated Britain and France, failed to conciliate the Soviet Union, and, after a brief liaison (1985–7), her relations with the United States were disrupted and soured by the latter's handling of the Iran–Contra affair. Thereafter the United States was viewed as a hostile power due to her intervention in the Gulf and her assistance to Kuwait. Only Israel gave covert aid to Iran, on the principle that both were opposed to Iraq.

There were paradoxical aspects to the conflict. For example, so far from widening the rift between the superpowers, their similar views actually promoted co-operation at the United Nations. Secondly, it had been widely feared that prolonged conflict between two major oil-producing countries would cause a disastrous shortage. Yet although their production *was* greatly reduced and war raged in the Gulf supply routes, the world was able to replace these losses and prices actually fell. The net result was to reduce the Gulf region's annual output by about one-third. Although oil provided an obvious material cause of conflict, the war seems to have been primarily about prestige, with neither of the authoritarian leaders willing to compromise. Consequently the greatest conflict in the region since the Second World War had an almost completely negative outcome: it resulted in no significant boundary changes; and no alteration in the state systems or their regimes.[37]

Finally, the Gulf War in 1990–1 provided a clear case of flagrant aggression by a notoriously repressive and barbarous regime which secured an unusual degree of consensus by the United Nations (with even Russia concurring) to expel the Iraqi forces from Kuwait. Although the United States and its principal allies were potentially vastly superior to Iraq it was not so obvious how that power could be brought to bear against the Iraqi defences in and around Kuwait. The sensibilities of hostile or non-cooperative Arab states could not be ignored, and there was a real danger that Israel would be drawn in. In the event, outstanding achievements in logistics and in air power prepared the way for the culminating land-air operations which lasted exactly 100 hours.

Only Saddam Hussein and his captive news media could deny that the UN forces had won a shattering victory, first in destroying installations within Iraq and then in annihilating large elements of the enemy ground forces in the battle for Kuwait and in the disorderly retreat. Indeed, it seems that President Bush called a sudden halt to the pursuit precisely because

Western liberal consciences were offended by what was becoming a massacre of the beaten army—the 'turkey-shoot' on the Basra road. But, as
Freedman and Karsh point out:

In military terms it was hard to know what the fuss was about. In both the theory
and practice of war this was the ideal opportunity to pursue and destroy a beaten
enemy. At the time General Schwarzkopf noted that it would be easy to get to
Baghdad, while Sir Peter de la Billière and other field commanders, as well as the
Saudis, were unhappy about the sudden ceasefire. True, only seven out of forty-
three Iraqi divisions were then deemed capable of fighting, but about 700 of their
tanks were undestroyed.

   The US President was, however, observing to the letter the UN mandate
which permitted the liberation of Kuwait but made no reference to the
invasion of Iraq. Nevertheless, it was the unofficial but widely assumed
Allied war aim that an overwhelming military defeat would lead to the flight
or enforced removal of Saddam Hussein. Not only did this not happen, but
within a few months the defeated dictator was publicly proclaiming that
Iraq had won a great victory against more than thirty nations led by the
United States.[38] His brutal repression of the Kurdish uprising caused a
sickening sense of failure in the West, and in particular tarnished President
Bush's reputation, due to his apparent encouragement of the Kurds during
the brief euphoria of the successful campaign.
   Consequently, although Iraq's defeat is still manifest in the economic and
arms penalties imposed on her, and in the continued American pressure to
enforce a 'no-fly zone' in the north to protect the Kurds (and in the south,
in an inadequate effort to protect the Marsh Arabs against Saddam's perse-
cution), much discussion in the West has revolved around the question as to
whether the anticipated benefits of military victory were not largely thrown
away by a premature ceasefire. Whether or not one finds them persuasive,
there were several reasons for the President's decision, in addition to the
humanitarian one already mentioned. Bush wished to obviate—or at any
rate minimize—charges of American aggression from the Arab world,
which would have been greatly increased by advancing any further into
Iraq. Such a move also risked alienating his Arab allies in the campaign,
notably Saudi Arabia and Egypt. Even Israel was content that Saddam
should remain in power; indeed, the only state in the region anxious to
remove him was Syria. There was a very real danger of the United States
becoming responsible for maintaining order in Iraq and of failing to find a

successor to Saddam Hussein who could command popular support. Lastly, there was the more Machiavellian concern with the post-war balance of power in the region as between Iran and Iraq, both of whom over recent years had been the recipients of large-scale Western arms sales. Ironically, had Saddam inflicted heavy losses on the Coalition forces he would probably have been toppled; the one-sided battle culminating in the 'slaughter of hapless Iraqis' enabled him to survive.[39]

Against all these substantial points, however, one must set Saddam Hussein's appalling record of human-rights violations towards his own religious and ethnic minorities, which has continued to the present as though the military defeat never occurred, and the wanton destruction and brutality that marked his brief occupation of Kuwait. In this respect securing a precise mandate from the United Nations and observing it to the letter has imposed a terrible handicap on those who wished to see Saddam Hussein overthrown and a more responsible government set up in his place. It is rather as though the Western allies had been compelled to halt their advance on the Rhine in March 1945 (and the Russians on the Oder-Neisse Line), leaving Hitler and the Nazis in power. The ironic result of the Gulf War seems to be that either Saddam Hussein will be left in power to build up his forces for renewed aggression or, by some means short of another great coalition war, he will have to be deposed. There can rarely have been a case in history where the chasm between a decisive military victory and an unsatisfactory political outcome has been so wide. It was a 'triumph without victory'.

It would clearly be premature to draw any but tentative conclusions about the character of conventional war in the next few years.[40] Despite the more widespread threats to the internal stability of states and to international order caused by a variety of low-intensity conflicts below the level of conventional wars, there have been enough of the latter to cast doubt on John Keegan's optimistic prophecy that conventional battle has simply become too terrible to fight. Indirect pressure or outright intervention by the superpowers or the United Nations have frequently occurred to bring conventional wars to a speedy and premature ceasefire before either side could secure a decisive victory. On the other hand, even small and weak states such as Somalia and Serbia can defy the United Nations by plunging their own and neighbouring people into anarchic civil war. Some stronger states, such as Israel, do not hesitate to employ force to defend their frontiers, and vital interests. We may conclude that the notion of 'war as an instrument of policy' and the pursuit of victory in battle now have little attraction for a large number of states, notably the liberal democracies, but the fate of Lebanon offers a terrible example of what may happen to a state which cannot defend itself against internal and external threats. Moreover, although Kuwait was successfully liberated, it is doubtful whether many future conflicts will have principles so clear-cut, aggressors so willing to take on the United Nations' military power on the latters' terms, and circumstances so favourable to a quick and militarily decisive victory.[41]

∾ *Conclusion* ∾

'I am tired and sick of war. Its glory is all moonshine . . . War is hell.' Thus spoke General William Tecumseh Sherman to the cadets of the Michigan Military Academy on 19 June 1879.[1] As the principal scourge of the Confederacy, Sherman knew better than most that the butcher's bill of battles and the physical destruction were only the first and obvious tragedies of war; in the longer term, wars leave a legacy of privation, anguish, suffering, and bitterness—especially in the defeated nation. Most servicemen who have experienced combat would endorse Sherman's sentiment: there are few militarists in the trenches, or among professional soldiers in general. Military historians also, though fascinated by this most dramatic and exciting of human group activities, are rarely captivated by war itself. Quite apart from the brutality and random suffering involved, they know that war is at best a blunt instrument of policy for resolving international disputes; that victory always has its price; and that its benefits may be disappointing or even illusory.

It is nevertheless important to underline the truth of common, everyday experience that if victory is often costly, defeat is always worse. Defeat in war in particular may bring unexpected compensations, such as economic resuscitation at the hands of the victors, or a political transformation from an authoritarian to a more democratic regime. But more often defeat has entailed physical devastation, military occupation, expulsion of population, punitive peace terms, and, perhaps worst of all, long-lasting humiliation and internal hatreds.

It would be reassuring if the foregoing remarks were too platitudinous to be worth stating but, as Edward Luttwak has sharply reminded us: 'The West has become comfortably habituated to defeat. Victory is viewed with great suspicion, if not outright hostility. After all, if the right-thinking are to achieve their great aim of abolishing war they must first persuade us that victory is futile or, better still actually harmful.'[2] Luttwak's strictures apply mainly to the United States, but similar tendencies are also evident in some sections of the British media. A reviewer of a political biography neatly captured this quirky, or even inverted, sense of values in remarking: 'The British will forgive defeat, but never victory.' To some extent this may be the fault of professional historians for failing to make the results of research, and hence of changing perspectives, available to the general public. Serious historians of the First World War, for example, have long since moved on to more complex issues than the stereotypes of 'butchers and bunglers' which still seem to obsess the authors and producers of anniversary programmes.

Perhaps the approach to war is also strongly influenced by a generation gulf between the middle-aged and elderly who experienced the Second World War, or were at least old enough to be aware of it, and those to whom it is as remote as the Boxer Rebellion or the Relief of Mafeking.[3] For the former, now at least in their late fifties, the Second World War posed no fundamental political or ethical dilemmas. Britain's role in the long struggle to defeat Nazi Germany was an honourable, even at times glorious one under the direction of an inspiring war leader in Winston Churchill. Revelations of Nazi barbarity in the latter part of the war, and in the post-war Nuremberg trials, added the retrospective justification of a moral crusade against evil to what had initially been primarily a continuation of the 'Thirty Years War' to prevent the German domination of Europe. Even the radical historian A. J. P. Taylor was moved to describe it without qualification as 'a good war'.

Contrast this comparatively clear, 'black and white' picture of Britain and her allies securing a hard-won victory in a necessary war against an evil regime, with the understanding of the post-1945 generations growing up under the threat of a possible nuclear holocaust with the former allies conducting a 'cold war'; and their successors witnessing the United States' humiliating defeat in Vietnam as their exemplar in conventional or guerrilla warfare. The complex issues raised by twentieth-century warfare are often treated in a cavalier way by revisionists who tend to dwell on their own nation's shortcomings or propose unrealistic alternatives to the decisions actually taken. These include, to mention just a few examples: that Britain could and should have remained neutral in 1914; that the war was fought for no political purpose; its conduct by the generals was criminally incompetent and its outcome futile—or worse since it served to set the stage for the Second World War. Britain should have stayed out of that war too, or, having made a blunder in 1939, should have made peace with Hitler in

1940. Churchill has been portrayed by revisionists as a militarist who sought victory for its own sake and in doing so lost the British Empire. Bomber Command is said to have deliberately waged war against 'innocent women and children', and wantonly destroyed Dresden when the war was virtually over. The Allied policy of 'unconditional surrender' was, to the 'right thinkers', indefensible because it stiffened German civilian determination and prevented the West from giving encouragement to the anti-Nazi resisters, thereby prolonging the war. In doing this it permitted the Soviet Union to extend its authoritarian rule deep into Central and south-eastern Europe, thus replacing one tyranny with another equally evil, so rendering the hard-won military victory hollow and meaningless.

For Luttwak's 'right thinkers', the culmination of the West's immoral conduct of the Second World War was the dropping of the atomic bombs on Japan, even though that beleaguered nation was in the process of surrendering. The list of Western 'errors' and 'atrocities' could be greatly extended by covering more recent conflicts, but just one episode may be cited as representative; namely, the prolonged public agonizing and controversy chiefly inspired by the Labour MP, Tam Dalyell, which followed the British government's decision to sink the Argentinian warship *Belgrano*, even though it was reportedly moving away from the battle zone.

What these and numerous other media controversies seem to signify is an undisguised zeal in disputing, decrying, or ridiculing Allied military successes, and an unwillingness to accept that one's own side might also have to resort to ruthless tactics in order not to lose a 'total war'—or be humiliated in a minor conflict. Here is the nub of the problem in that most of the strictures (such as those concerning the bombing or Dresden and of Hiroshima and Nagasaki) rest on the firm foundation of hindsight that the war was eventually won with consequences which are also familiar to the critics. If this study causes some of these controversies to be reappraised then it will have performed a useful function.

My aim has been to place the pursuit of victory—and the avoidance of defeat—in historical context from the mid-eighteenth century to the present. I have paid careful attention to two related but in fact often ill-coordinated problems. First, the operational difficulties involved in winning clear, 'decisive' victories in the battle or campaign; and, secondly, the immense and perpetual problems of imposing political control throughout and beyond operations into the peace negotiations, with a view to rendering

warfare an effective instrument of state or alliance policy. Political control
depends of course on several variables: most obviously the statesman's
ability to impose his will on his generals; but also his capacity to heed and
accommodate political concerns of the loser and, extremely important,
those of other interested parties. If this study rebuts the extreme view that
war is 'always futile' and 'never pays', it also suggests that, even with an
unmatched combination, such as that of the elder Moltke and Bismarck,
military victory never provides a perfect solution. In principle, no military
defeat is final and the victor does well to bear this in mind in the conditions
he imposes.

One of the main arguments in this book is that the French Revolution
and, more specifically, Napoleon *did* introduce a more dramatic and
destabilizing mode of warfare, which sought the utter destruction of the
enemy's army as a prelude to the ruthless subordination of his state to
French interests. Indeed, as Paul W. Schroeder's brilliant analysis stresses,
Napoleon attempted, though in vain, to convert the whole of Europe into a
colonial dependency.[4] Napoleon's eventual military defeat and nemesis, so
far from eclipsing his legend of conquest and glory, seems actually to have
contributed to the Napoleonic 'legacy' which, with its emphasis on the
pursuit of victory through decisive battle, dominated military thinking up to
the First World War. In the aftermath of that tragically prolonged and
attritional struggle it became conventional wisdom that the Napoleonic
pursuit of decisive victory had been a snare and a delusion. Certainly,
military, political, and economic developments in the nineteenth century
suggested to most military thinkers that it would be extremely difficult to
repeat such dramatic victories as Ulm, Austerlitz, or Waterloo in changing
conditions, while anti-war crusaders were convinced that it had become
impossible.

Unfortunately for the latter, Prussia's remarkably quick, inexpensive, and
politically profitable victories in 1866 and 1870–1 suggested that the
Napoleonic model could still be made to work and, furthermore, that war
remained an effective instrument of state policy. Japan's spectacular sea and
land victories over Russia in 1904–5 provided further ammunition for those
who believed that Napoleonic principles could still be successfully implemented in an era of railways, machine-guns, and mass conscription.

Such optimism was badly shaken by the length and nature of the First
World War, but even in this case the historian must be wary of arguing that

military conditions made stalemate inevitable; far more influential in fact
was the number of 'players' and the statesmen's loss of control over their
military mastodons. The First World War was widely believed to have
signalled the end of an era—indeed a millenium—of a European-dominated states system based on a balance of power frequently adjusted by
inter-state wars. Yet at the same time it was abundantly clear to the general
public, and not just to military experts, that technological development,
especially in air power and in armoured vehicles, was once again rendering
military operations more mobile and potentially decisive.

This was demonstrated, on a global scale, by the blitzkrieg conquests of
Italy, Germany, and Japan between the years 1939 and 1942. The emphasis
now was on decisive campaigns rather than set-piece battles, but the net
result was the same: crushing military victories led to the subordination or
virtual extinction of European states and the collapse of the European
empires in the Far East. These conquests may be seen as embodying the
concept of 'war as an instrument of policy', but political tolerance and
moderation to make the victories acceptable to the vanquished (let alone
the unvanquished) were conspicuously lacking. Indeed, I have argued that
the very resort to brutal, aggressive conquests in the manner of Germany
and Japan was incompatible with Clausewitz's notion of war as an instrument of policy. Rather they approximated to Napoleon's style of total
conquest, entailing defiance of existing international conventions and the
complete destruction of the existing states system.

Since 1945 the employment of war as an instrument of state policy
has not disappeared, but it has been markedly confined and restricted as
between major powers, not only by the fear of nuclear escalation, but also
by the constraints imposed by alliances and international organizations
and, encouragingly, by the complete acceptance in large areas of the world,
that war would be utterly counter-productive in furthering national interests. However, the 'pursuit of victory' has not been abandoned everywhere,
and a number of case-studies are discussed in the final chapter. Since
the Falklands War in 1982 'triumphalism' has been widely deplored in
Western democracies, and the celebration of victory in the Gulf conflict
was more subdued—in part no doubt because the victory itself was so
unsatisfactory.

Here surely lies an important pointer to the likely frequency and nature
of future conflicts. With a few possible exceptions, major powers are un-

likely in the foreseeable future to resort to war unilaterally in pursuit of national interests. Rather they will fight as members of international coalitions, with all the restraints on war aims, conduct of operations, and peace terms such delicate collective enterprises entail. In these conditions the pursuit of victory is likely to be hedged about with restrictions which Napoleon would have regarded as intolerable. However, in the collective resistance to aggression, as exemplified by Saddam Hussein's attempted annexation of Kuwait, striving for military victory under close political direction will be as important as ever.

NOTES

Introduction

1. R. F. Weigley, *The Age of Battles: The Quest for Decisive Warfare from Breitenfeld to Waterloo* (1991).
2. B. H. Liddell Hart, *The Ghost of Napoleon* (1934). J. Colin, *Transformations of War* (1912). Azar Gat, *The Development of Military Thought: The Nineteenth Century* (1992).
3. Jay Luvaas, *The Military Legacy of the Civil War: The European Inheritance* (1959).
4. Michael Howard, 'The Forgotten Dimensions of Strategy', in *The Causes of Wars* (1983), 103.
5. For a recent reassessment of Grant's skill as an operational commander see Brian Holden Reid, 'Another Look at Grant's Crossing of the James, 1864', *Civil War History*, 39/4 (Dec. 1993), 291–316. See esp. 294, 297–8, 314, for the pitfalls of judging Grant's generalship anachronistically on 20th-cent. criteria.
6. D. C. Watt, *Too Serious a Business: European Armed Forces and the Approach to the Second World War* (1975).
7. Allan R. Millett and Williamson Murray (eds.), *Military Effectiveness*, iii. *The Second World War* (Allen & Unwin, 1988), chs. 1 and 5.
8. For a recent example of this wisdom in hindsight see István Deák, 'Misjudgement at Nuremburg', *New York Review of Books* (7 Oct. 1993) and the resulting correspondence, ibid. (13 Jan. 1994).

Chapter 1

1. Peter Paret (ed.), *Makers of Modern Strategy from Machiavelli to the Nuclear Age* (1986), chs. 4–7. Spenser Wilkinson, *The French Army before Napoleon* (1991). Azar Gat, *The Origins of Military Thought from the Enlightenment to Clausewitz*(1989), chs. 2–5.
2. J. Colin, *The Transformations of War* (1912), 221.
3. R. R. Palmer, 'Frederick the Great, Guibert, Bülow: From Dynastic to National War' in Paret, *Makers*, 91–2.
4. Carl von Clausewitz, *On War*, ed. M. Howard and P. Paret (1976), 590.
5. Ibid. 590–1.
6. Ibid. 179.
7. J. P. Sainte-Etienne quoted in M. S. Anderson, *War and Society in Europe of the Old Regime* (Fontana Paperback, 1988), 189.

8. Hew Strachan, *European Armies and the Conduct of War* (1983), 14–15. Jeremy Black, 'Eighteenth-Century Warfare Reconsidered', *War in History*, 1/2 (1994), 215–32.

9. Colin, *Transformations*, 198–201. See also John Keegan, *A History of Warfare* (1993), 344–5.

10. Palmer, 'Frederick', 95. Gerhard Ritter, *Frederick the Great* (1968), 130. Christopher Duffy, *Frederick the Great: A Military Life* (1985), 292–6. Black, *War in History*, 217, makes a good case for studying 18th-cent. warfare in its own terms rather than as a mere prologue to the 'real' war of the Napoleonic era.

11. *On War*, 590–3.

12. Colin, *Transformations*, 177, suggests a steady diminution of the percentage losses of killed and wounded from an average of 25–30% in the 18th cent. to 20% in the Napoleonic Wars and 10% in the Franco-Prussian War up to 4 Sept. 1870. Colin thought this was mainly due to more prisoners being taken, but from the later 19th cent. improved medical care was at least as important. Black, *War in History*, 218 ff. shows that 18th-cent. generals persistently aimed at decisive victory but were often thwarted by physical conditions.

13. Weigley, *The Age of Battles* (1991), 182–92. See also Duffy, *Frederick the Great*, 114–20.

14. Palmer, 'Frederick', 94. Jay Luvaas, *Frederick the Great on the Art of War* (1966), 139.

15. Luvaas, *Art of War*, 139 ff., quoting Frederick on 'The Anatomy of Battle'.

16. Duffy, *Frederick the Great*, 302. Also see the map in Weigley, *Age of Battles*, 169.

17. Ritter, *Frederick the Great*, 135–40. Palmer, 'Frederick', 104.

18. *On War*, 260.

19. Weigley, *Age of Battles*, pp xii–xiii; emphasis added.

20. A. J. Bacevich, *Parameters* (Summer 1993), 111–13.

21. Weigley, *Age of Battles*, 171–6.

22. Ritter, *Frederick the Great*, 120–2. Luvaas, *Art of War*, 22.

23. Duffy, *Frederick the Great*, 227–42. Weigley, *Age of Battles*, 192.

24. Duffy, *Frederick the Great*, 227–42. Weigley, *Age of Battles*, 185–93. Ritter, *Frederick the Great*, 126–7.

25. Duffy, *Frederick the Great*, 284–5. Gordon A. Craig, 'Delbrück: The Military Historian', in Paret (ed.), *Modern Strategy*, 342–3. The German terms employed by Delbrück were *Niederwerfungsstrategie* (annihilation) and *Ermattungsstrategie* (exhaustion).

26. Ritter, *Frederick the Great*, 130. Luvaas, *Art of War*, 22–3.

27. Ibid. 23 ff.

28. Ibid. 10–11, 25. I am grateful to Christopher Duffy for pointing out that Frederick's final campaign was still aggressive and offensive in intention. See

C. Duffy, *The Army of Frederick the Great* (1974), 204–5. See also Duffy, *Frederick the Great*, 267–78.

29. Luvaas, *Art of War*, 25, 359–60.
30. Duffy, *Frederick the Great*, 304. Palmer, 'Frederick', 96–8.
31. Ritter, *Frederick the Great*, 128. Duffy, *Frederick the Great*, 322–3.
32. Colin, *Transformations*, 207, 211.

Chapter 2

1. M. Howard, *War in European History* (1976), 75–6.
2. P. Paret, 'Napoleon and the Revolution in Warfare', in P. Paret (ed.), *Makers of Modern Strategy from Machiavelli to the Nuclear Age* (1986), 124–6.
3. Ibid.
4. J. Colin, *The Transformations of War* (1912). S. Wilkinson, *The French Army before Napoleon* (1991) and *The Rise of General Bonaparte* (1991).
5. Wilkinson, *French Army*, 20–2.
6. Paret, *Makers*, 124–5.
7. Colin, *Transformations*, 210–14.
8. M. Van Creveld, *Command in War* (1985), 63.
9. Ibid. 64.
10. Gunther E. Rothenberg, *The Art of Warfare in the Age of Napoleon* (1977), 147. Colin, *Transformations*, 228–95 *passim*.
11. Rothenberg, *Art of Wafare*, 149.
12. R. F. Weigley, *The Age of Battles* (1991), 378–90. J. F. C. Fuller, *The Conduct of War 1789–1961* (1972), 44–52.
13. Weigley, *Age of Battles*, 389.
14. Ibid. 390–8. Paret, *Makers*, 132.
15. Weigley, *Age of Battles*, 418–34.
16. Paret, *Makers*, 134.
17. D. Kaiser, *Politics and War: European Conflict from Philip II to Hitler* (1990), 237–47.
18. Ibid. 248–9.
19. Rothenberg, *Art of Warfare*, 54. The non-French majority included Dutch, Westphalian, Polish, Bavarian, Saxon, Prussian, Austrian, Swiss, Italian, and even Spanish troops.
20. Howard, *European History*, 84–5. Weigley, *Age of Battles*, 482. Rothenberg, *Art of Warfare*, 134, notes that between 1800 and 1815 more than $2\frac{1}{2}$ million French citizens were mustered but only $1\frac{1}{2}$ million were actually enlisted—less than 7% of the population of France proper.
21. Paret, *Makers*, 137. Fuller, *Conduct*, 53–5.
22. Paret, *Makers*, 137.
23. Rothenberg, *Art of Warfare*, ch. 6 *passim*. See also P. Paret, *Yorck and the Era*

of Prussian Reform (Princeton: Princeton UP, 1966), and Richard Glover, *Peninsular Preparation: The Reform of the British Army 1795–1809* (Cambridge: CUP, 1963).

24. Wilkinson, *French Army*, 12.
25. Pieter Geyl, *Napoleon: For and Against* (Cape, 1949), 325, quoting a remark made by P. Muret in 1913. See also Kaiser, *Politics*, 245.
26. Paret, *Makers*, 136.
27. Ibid. 137–8. Kaiser, *Politics*, 260–3.
28. Paul W. Schroeder, 'Napoleon's Foreign Policy: A Criminal Enterprise', *Journal of Military History*, 54/2 (Apr. 1990), 147–61. See also his *The Transformation of European Politics, 1763–1848* (1994), 276–86.
29. Schroeder, 'Napoleon's Foreign Policy', 155.
30. Ibid. 159. 'They were used to high-stakes poker . . . but now they discovered that the game was run by someone who always cheated, held the biggest guns as well as the high cards, made his own rules, always won, and never paid off.'
31. Kaiser, *Politics*, 255, depicts British policy as just as ruthless and selfish as Napoleon's—but more realistic and successful.
32. Rothenberg, *Art of Warfare*, app. 1, lists 78 battles between 1792 and 1815 of which Napoleon personally commanded in 34, winning 28 and losing 6 (but counting Quatre Bras, Ligny, and Waterloo as 3)—an impressive record but surely not of much consolation to him on St Helena.
33. Paret, *Makers*, 138–42.

Chapter 3

1. Azar Gat, *The Development of Military Thought: The Nineteenth Century* (1992), 2–3, and *The Origins of Military Thought from the Enlightenment to Clausewitz* (1989), 203–4.
2. Antoine-Henri de Jomini, *The Art of War* (1992). This is a repr. of the 1862 edn. with a new introduction by Charles Messenger. Gat, *Origins*, 115–17. John Shy, 'Jomini', in P. Paret (ed.), *Makers of Modern Strategy from Machiavelli to the Nuclear Age* (1986), 151–3.
3. M. Howard, *Clausewitz* (1983), 45–6.
4. C. von Clausewitz, *On War*, ed. M. Howard and P. Paret (1976), 258.
5. Gat, *Origins*, 206, 211. Jomini, *Art of War*, 252–77.
6. Gat, *Origins*, 132–5.
7. Ibid. 113. Shy, 'Jomini', 144. Jomini, *Art of War*, ch. 3 *passim*.
8. Shy, 'Jomini', 145–6.
9. Gat, *Origins*, 122.
10. Ibid. 119–21. Jomini, *Art of War*, 130. For a lively critical analysis of strategic concepts see Edward N. Luttwak, *Strategy: The Logic of War and Peace* (Cambridge, Mass.: Harvard UP, 1987).

11. Jomini, *Art of War*, 70–1, 85–92, 334. Shy, 'Jomini', 173–5.

12. Jomini, *Art of War*, 48–9, 137–8, 321–5.

13. L. M. Crowell, 'The Illusion of the Decisive Napoleonic Victory', *Defense Analysis*, 4/4 (Dec. 1988), 329–46. David R. Jones, 'The Napoleonic Paradigm: The Myth of the Offensive in Soviet and Western Military Thought', in David A. Charters, Marc Milner, and J. Brent Wilson (eds.), *Military History and the Military Profession* (Westport, Conn.: Praeger, 1992), 211–27.

14. Shy, 'Jomini', 153–5.

15. Ibid. 153, 169–71. Jomini, *Art of War*, 34–5.

16. Gat, *Development*, 20–3, 43. The phrase about Jomini and Civil War generals was coined by J. D. Hittle. Alfred Thayer Mahan, Denis's son, was perhaps an even more influential Jominian.

17. Shy, 'Jomini', 178–80.

18. Jehuda L. Wallach, 'Misperceptions of Clausewitz's *On War* by the German Military', *Journal of Strategic Studies*, 9 (June/Sept. 1986), 220–1: 'Essentially it was a Jominian rather than a Clausewitzian attitude that dominated [19th-cent. military] thinking.' P. Paret, 'Clausewitz and the Nineteenth Century', in M. Howard (ed.), *The Theory and Practice of War* (Cassell, 1965), 31.

19. A notable exception is John Keegan. See the frequent critical refs. to Clausewitz in *A History of Warfare* (1993).

20. In addition to works already cited above see e.g. P. Paret, *Clausewitz and the State* (1985) and W. B. Gallie, *Philosophers of Peace and War* (1978), 37–65.

21. Gat, *Origins*, 199.

22. Howard, *Clausewitz*, 46. *On War*, 259.

23. Gat, *Origins*, 204–5, cf. Howard, *Clausewitz*, 47. Also relevant is Jan Willem Honig, 'The European Origins of the Concepts of "Limited" and "Total" War' (TS, awaiting publication).

24. Gat, *Origins*, 255–63.

25. *On War*, 501.

26. Ibid. 488–9. Gat, *Origins*, 213–14.

27. Ibid. 218–20.

28. Ibid. 221. *On War*, 605–6. Gallie, *Philosophers*, 61, warns against placing too much emphasis on Clausewitz as a political theorist. '*On War* is emphatically about war, and was primarily written for military men.'

29. Gat, *Origins*, 222–5. *On War*, 75.

30. Ibid. 235–6. Gallie, *Philosophers*, 60, proposes a reconstruction of Clausewitz's conceptual system.

31. See e.g. M. Howard's introduction to *On War*, 39–41, and Honig, 'Concepts', 9–10, who notes a tendency among modern Clausewitz interpreters to put too much retrospective emphasis on the concept of 'limited war' and the relationship between war and politics.

32. *On War*, 192.

33. Gallie, *Philosophers*, 61. See also C. B. A. Behrens's review of Paret's *Clausewitz and the State*, 'Which Side was Clausewitz On?', *New York Review of Books* (14 Oct. 1976).

34. Gat, *Origins*, 249–50.

35. Ibid. 129. Christopher Bassford, *Clausewitz in English: The Reception of Clausewitz in Britain and America, 1815–1945* (1994) demonstrates that Clausewitz's works were somewhat better-known than is widely assumed. But his intelligent discussion also shows why it remains extremely difficult to be precise about Clausewitz's influence.

Chapter 4

1. Kalevi J. Holsti, *Peace and War: Armed conflicts and international order, 1648–1989* (1991), 140–3. Holsti's list of wars and issues between 1815 and 1914 suggests that the international order constructed in 1814–15 was much more successful than previous attempts in 1648 and 1713. See also Paul W. Schroeder, *The Transformation of European Politics, 1763–1848* (1994), 575–82.

2. Michael Howard, *War in European History* (Oxford: 1976), 94.

3. John Keegan, *The Face of Battle* (1976), 57–62. Anglo-Saxon culture in the 19th cent. was paradoxically both martial yet anti-military, see Marcus Cunliffe, *Soldiers and Civilians: The Martial Spirit in America, 1775–1865* (Eyre & Spottiswoode, 1969).

4. Sir Edward Creasy, *Fifteen Decisive Battles of the World* (Everyman paperback edn., 1963), 418.

5. B. H. Liddell Hart, *Decisive Wars of History* (Bell, 1929). J. F. C. Fuller, *Decisive Battles: Their Influence upon History and Civilization* (2 vols. Eyre & Spottiswoode, 1939–40) and *The Decisive Battles of the United States* (Hutchinson, 1942). See also Brian Holden Reid, 'Theory from Practice: Major-General J. F. C. Fuller', *History Today* (June 1989), 44–9.

6. Keegan, *Face*, 62 and see also Dennis E. Showalter, 'Of Decisive Battles and Intellectual Fashions: Sir Edward Creasy Revisited', *Military Affairs*, 52/4 (Oct. 1988), 206–8.

7. William Carr, *The Origins of the Wars of German Unification* (1991), ch. 3.

8. A. J. P. Taylor, *The Struggle for Mastery in Europe, 1848–1918* (1969), 164–6.

9. Carr, *Origins*, 136–8; Gordon A. Craig, *The Battle of Königgrätz* (1965), 11–39; Brian Bond, 'The Austro-Prussian War', *History Today* (Aug. 1966), 538–46.

10. Hew Strachan, *European Armies and the Conduct of War* (1983), 98–101.

11. Jean Colin, *The Transformations of War* (1912), 138–41, 321, 315–16.

12. Craig, *Königgrätz*, 84–98.

13. Ibid. 176–8.

14. Ibid. 185. Colin, *Transformations*, 349.

15. Craig, *Königgrätz*, 186.

16. Gordon A. Craig, *The Politics of the Prussian Army 1640–1945* (New York: paperback edn., 1964), 193–204. See also Gerhard Ritter, *The Sword and the Scepter*, i. *The Prussian Tradition* (1969), 187–216. Ritter takes a much more favourable view of Moltke's attitude to civil–military relations than Craig.

17. Craig, *Politics*, 199–200.

18. Ibid. 202–3. Carr, *Origins*, 138–9.

19. Taylor, *Struggle*, 168–9. Craig, *Politics*, 203–4.

20. Gordon A. Craig, *Germany: 1866–1945* (1978), 8–14. Taylor, *Struggle*, 210.

21. Sherman's remarks are quoted in Strachan, *European Armies*, 74–5. On Lincoln's aim in 1861 and 1862 to combine limited force with a policy of conciliation, see Russell F. Weigley, 'The American Military and the Principle of Civilian Control', *Journal of Military History*, 57/5 (Oct. 1993), 32–4.

22. See esp. Jay Luvaas, *The Military Legacy of the Civil War: The European Inheritance* (1959). Brian Holden Reid, 'British Military Intellectuals and the American Civil War', in C. Wrigley (ed.), *Warfare, Diplomacy and Politics* (Hamish Hamilton, 1986). See also Brian Bond, *Liddell Hart: A Study of his Military Thought* (1977), 47–9, for the keen publishing interest in Civil War military biographies in the 1920s.

23. G. F. R. Henderson, *The Campaign of Fredericksburg* (1886), repr. in Jay Luvaas (ed.), *The Civil War: A Soldier's View* (Chicago: Chicago UP, 1958), 9–119. Richard Holmes, *The Road to Sedan* (London: Royal Historical Society, 1984), 199–233. Strachan, *European Armies*, 60–75, 101. See also Brian Holden Reid, 'Another Look at Grant's Crossing of the James, 1864', *Civil War History*, 39/4 (1993), for criticism of the tendency to exaggerate the 'modernity' of the Civil War.

24. Holsti, *Peace and War*, 60–75, 101.

25. Taylor, *Struggle*, 204.

26. Craig, *Germany*, 27.

27. Ibid. 28.

28. Colin, *Transformations*, 334, points out that France in 1870 made the basic error of assembling her forces in successive waves rather than altogether. Thus after the loss of some 300,000 in the battles up to and including Sedan, there were still nearly the same number of reservists at the depots and a further 100,000 *gardes mobiles*. But for this mistake France could have put as many as 700,000 troops into the field by 1 Sept.

29. The standard history of the conflict in English remains Michael Howard, *The Franco-Prussian War* (1967). See also Strachan, *European Armies*, 101–3, and Brian Bond, *War and Society in Europe, 1870–1970* (1984), 15–21.

30. Craig, *Politics*, 206–7.

31. Stig Förster, 'Facing "Peoples' War": Moltke the Elder and Germany's Military Options after 1871', *Journal of Strategic Studies*, 10/2 (June 1987), 209–30.

32. Ibid. 215.

33. Craig, *Politics*, 204–5.

34. Ibid. 208–9.

35. Ibid. 211.

36. Ibid. 212–14.

37. Carr, *Origins*, 206.

38. Taylor, *Struggle*, 211, 217–18. Craig, *Germany*, 30.

39. Ibid. 34.

40. Förster, 'Facing "People's War" ', 217.

41. Ibid. 220.

42. Ibid. 221.

43. Ibid. 223–4.

44. Dennis E. Showalter, 'Total War for Limited Objectives: An Interpretation of German Grand Strategy', in Paul Kennedy (ed.), *Grand Strategies in War and Peace* (1991), 105–24.

45. 'In Moltke's operations of 1866 and 1870 we find the strategy which would naturally be produced by the preaching [*sic*] of Clausewitz.' Colin, *Transformations*, 300–3.

Chapter 5

1. See e.g. Jack Snyder, *The Ideology of the Offensive* (Ithaca, NY: Cornell UP, 1984). Stephen Van Evera, 'The Cult of the Offensive and the Origins of the First World War', *International Security* (Summer 1984), 58–107.

2. Gideon Y. Akavia, *Decisive Victory and the 'Cult of the Correct Military Doctrine': The Case of French Military Doctrine before 1914* (Haifa: CEMA, Apr. 1991).

3. Wolfgang J. Mommsen, 'The Topos of Inevitable War in Germany in the Decade before 1914', in V. R. Berghahn and M. Kitchen (eds.), *Germany in the Age of Total War* (1981), 23–45.

4. Ibid. 31.

5. Ibid. 32.

6. Ibid. 38, 41. Gordon A. Craig, *Germany: 1866–1945* (1978), 319–20.

7. Brian Bond, *War and Society in Europe 1870–1970* (1984), ch. 3. Paul M. Kennedy, *The Rise of the Anglo-German Antagonism, 1860–1914* (Allen & Unwin, 1980).

8. Colmar von der Goltz, *The Nation in Arms* (1903), 386.

9. Ibid. 128–9. Bond, *War and Society*, 47.

10. Von der Goltz, *Nation*, 381–4.

11. Ibid. 385.

12. Ibid. 388–91.

13. F. von Bernhardi, *Germany and the Next War* (1912), 12.

14. Ibid. 156–7.

15. Ibid. 290.

16. Azar Gat, *The Development of Military Thought: The Nineteenth Century* (1992), 85.

17. I. F. Clarke, *Voices Prophesying War, 1763–1984* (1970), ch. 4 *passim*. I draw heavily in what follows on pp. 108–43.

18. Bond, *War and Society*, 83. On 13 Sept. 1914, after the battle of the Marne, a French officer asked General Sir Henry Wilson when he thought the Allied armies would cross into Germany. Wilson thought four weeks—the Frenchman three. Both erred on the optimistic side. See C. E. Callwell, *Field Marshal Sir Henry Wilson* (Cassell, 1927), i. 177.

19. Theodore Ropp, *War in the Modern World* (New York: Collier paperback edn., 1962), 222.

20. I. S. Bloch, *Is War Impossible?* (1991). See also T. H. E. Travers, 'Technology, Tactics, and Morale: Jean de Bloch, the Boer War, and British Military Theory, 1900–14', *Journal of Modern History*, 51 (June 1979), 264–86.

21. Gat, *Development*, 110.

22. Ibid. 110–11.

23. Bloch's main ideas are summarized in his conversation with W. T. Stead, published as an introduction to *Is War Impossible?*

24. Roger Chickering, *Imperial Germany and a World Without War: The Peace Movement and German Society, 1892–1914* (Princeton: Princeton UP, 1975), 387–92, 403–5. For the revival of British interest in Bloch after 1918, see B. H. Reid, *J. F. C. Fuller: Military Thinker* (1987), 69.

25. A. Bucholz, *Moltke, Schlieffen and Prussian War Planning* (1991), 156.

26. G. Ritter, *The Schlieffen Plan: Critique of a Myth* (Wolf, 1958).

27. Bucholz, *Moltke*, 209.

28. M. van Creveld, *Supplying War: Logistics from Wallenstein to Patton* (1978), 113–41.

29. Ritter, *Schlieffen Plan*, 70–1.

30. Bond, *War and Society*, 89–90. Ritter, *Schlieffen Plan*, 91.

31. Gat, *Development*, 98. Gat points out that Tirpitz's naval planning at the same time was similar in its disregard of likely political consequences, and was also in broad terms well known to German statesmen.

32. L. L. Farrar Jnr, *The Short War Illusion* (Santa Barbara, Calif.: Clio paperback, 1973), 8–11.

33. Douglas Porch, *The March to the Marne: The French Army, 1877–1914* (1981), 162–231. David B. Ralston, *The Army of the Republic* (Cambridge, Mass.: MIT Press, 1967), 319–76.

34. Gat, *Development*, 110. Keith Neilson, ' "That Dangerous and Difficult Enter-prise": British Military Thinking and the Russo-Japanese War', *War and Society*, 9/2 (Oct. 1991), 17–37.

35. Neilson, 'Dangerous and Difficult', 20–3. Bond, *War and Society*, 84.

36. Ibid. 84–5. Neilson, 'Dangerous and Difficult', 31, writes: 'Quite simply the lessons of the Russo-Japanese War were not accepted because they were not unambiguous.'

37. J. Colin, *The Transformations of War* (1912), 41–6: 'as impressions faded the heroism of military writers showed itself on paper in a firm determination to achieve the impossible', but 'Fire [power] with its brutal realism, . . . will blow away in smoke all these fine theories about the employment of the masses'.

38. Ibid. 53–61.

39. Ibid. 325–31.

40. Ibid. 343.

41. Ibid. 226.

42. Akavia, *Decisive Victory*, 84, 99, 108–9. While sympathetic to the French theorists Du Picq, Foch, and Grandmaison, whose ideas have been frequently misquoted or taken out of context, Dr Akavia points out that Foch and Grandmaison particularly *were* sometimes guilty of muddled thinking.

43. Porch *March*, esp. 73–133, 232–45.

44. Gat, *Development*, ch. 3 *passim*. Michael Howard, 'Europe on the Eve of the First World War', in R. J. W. Evans and H. P. von Strandmann (eds.), *The Coming of the First World War* (1988), 1–18, and 'Men against Fire: The Doctrine of the Offensive in 1914', in P. Paret (ed.), *Makers of Modern Strategy from Machiavelli to the Nuclear Age* (1986), 510–26.

45. Gat, *Development*, 139.

46. Howard, 'Men against Fire', 523–6.

47. Howard, 'Europe', 6–9. Craig, *Germany*, 359.

48. G. Ritter, *Sword and Scepter*, ii. 7–136, and *Schlieffen Plan*, 90–6. One of the least 'military' powers at this time, the USA, had one of the most narrowly professional armies.

49. L. M. Crowell, 'The Illusion of Decisive Napoleonic Victory', *Defense Analysis*, 4/4 (Dec. 1988), 339.

50. Craig, *Germany*, 339–41.

51. Quoted in Bond, *War and Society*, 96.

52. See the glossary of euphemisms in Paul Fussell, *The Great War and Modern Memory* (Oxford: OUP, 1975), 22.

53. Howard, 'Europe', 15–17.

54. Ibid. 12.

55. Farrar, *Short War Illusion*, 150.

56. Gat, *Development*, 113.

Chapter 6

1. See e.g. John Terraine, *To Win a War* (Sidgwick & Jackson, 1978); C. N. Barclay, *Armistice 1918* (1969); Barrie Pitt, *1918: The Last Act* (Cassell, 1962).
2. A. J. P. Taylor, *The Struggle for Mastery in Europe, 1899–1918* (1969), 538–9.
3. D. C. Watt, *A History of the World in the 20th Century*, i. *1899–1918* (1970), 280–1. See also G. Blainey, *The Causes of War* (1973), 224–5.
4. M. Howard, *Studies in War and Peace* (1970), 104.
5. Compare R. Prior and T. Wilson, *Command on the Western Front* (1992), with T. Travers, *The Killing Ground* (1987) and *How the War was Won* (1991).
6. M. Kitchen, *The Silent Dictatorship* (1976). J. Joll, *Europe since 1870* (1976), 206–8.
7. Howard, *Studies*, 104–5.
8. D. Stevenson, *The First World War and International Politics* (1988), p. v.
9. J.-J. Becker, *The Great War and the French People* (1985), 325.
10. Taylor, *Struggle*, 564–5. Watt, *History*, 306–10.
11. Taylor, *Struggle*, 538. J. M. Roberts, *Europe 1880–1945* (1976), 291.
12. Stevenson, *First World War*, 103–6.
13. Ibid. 138. Watt, *History*, 294–5.
14. Stevenson, *First World War*, 195. Watt, *History*, 136.
15. Marc Ferro, *The Great War 1914–18* (1973), 214. According to Roberts, *Europe*, 291, no civilian was present at the conference which discussed the decision to launch Ludendorff's final offensive.
16. Travers, *How the War was Won*, 50–90.
17. Ferro, *Great War*, 216. His final chapter is entitled 'The Illusions of Victory'.
18. Paul Guinn, *British Strategy and Politics 1914 to 1918* (Oxford, Clarendon Press, 1965), 308–9. I am indebted to Robin White for reminding me of this reference.
19. Benjamin Schwartz, 'Divided Attention: Britain's Perception of a German Threat to her Eastern Position in 1918', *Journal of Contemporary History*, 28/1 (Jan. 1993), 100–22.
20. Ibid. 113.
21. *Field Marshal Sir Henry Wilson: His Life and Diaries*, ed. Charles Callwell (1927), ii. 118–19. As late as 1 Aug. 1918 L. S. Amery warned Henry Wilson of the excessive price which the Allies might pay for striving too hard for victory in the West in *1919* (my emphasis); namely, that 'the Boche will consolidate in the East, where we leave him alone, while we gain a few villages in France and Flanders': *The Leo Amery Diaries*, i. *1896–1929*, ed. John Barnes and David Nicholson (Hutchinson, 1980), 227, 230–1, 234.
22. *Wilson*, ed. Callwell, 120–1.
23. *The Private Papers of Douglas Haig 1914–1919*, ed. R. Blake (1952), 326–40.
24. S. Weintraub, *A Stillness Heard Round the World* (1985), 210.

25. See Avner Offer, *The First World War: An Agrarian Interpretation* (Oxford: OUP, 1989).

26. Tim Travers in *How the War was Won* perhaps overstates the argument that the German army defeated itself in 1918. In the late summer and autumn the Allies still had to impose their will on a powerful adversary in strongly fortified positions. Cf. T. Wilson, *The Myriad Faces of War* (1986), 584. The statistics of German losses are taken from Barclay, *Armistice*, 86.

27. Ferro, *Great War*, 217–18. Wilson, *Myriad Faces*, 618.

28. The significance of Allenby's excellent intelligence is stressed in a letter in the *Times Literary Supplement* (30 Apr. 1993) by Anita Engle referring to her book *The Nili Spies* (Hogarth Press, 1959).

29. Wilson, *Myriad Faces*, 618–20.

30. 'The Somme' in the British historical consciousness or folk memory is synonymous with the costly offensive of 1916 (strictly speaking not fought on that river at all), whereas only specialist military historians are aware of the successful Allied advance south of the river in 1918.

31. Barclay, *Armistice*, 54–6, 82–3, and Wilson, *Myriad Faces*, 596–607, provide excellent summaries.

32. Barclay, *Armistice*, 62, 81–2. Watt, *History*, 329–30. R. Bessel, *Germany after the First World War* (1993), 69–90.

33. Alan Sharp, *The Versailles Settlement: Peacemaking in Paris, 1919* (1991), 128. Weintraub, *Stillness*, 387–91. German soldiers returning to their homeland were welcomed with banners proclaiming 'UNCONQUERED IN THE FIELD', while even the Social Democratic Party's official pamphlets encouraged the troops to believe that they had not been beaten, but had simply stopped fighting on the orders of the Reich's leaders.

34. Stevenson, *First World War*, 201–2. Watt, *History*, 310–11. For the breast-beating among British politicians, intellectuals, and literary figures about the severity of the terms imposed on Germany see A. Lentin, *Guilt at Versailles: Lloyd George and the Pre-History of Appeasement* (Methuen, University Paperback, 1985). Lentin concludes that 'the Appeasers' in effect renounced the victory of 1918 as well as the Treaty of Versailles.

35. K. J. Holsti, *Peace and War* (1991), 213–18. See also Alan Sharp's judicious conclusions, *Versailles*, 185–96.

36. Wilson, *Myriad Faces*, 835–7.

37. Ibid. 837–8.

38. Ibid. 843–5.

39. F. H. Hinsley, *Power and the Pursuit of Peace* (1963), 309–22.

40. Wilson, *Myriad Faces*, 846–8.

41. This section draws heavily on Weintraub, *Stillness*, esp. 229–40, 253–63, 288, 315.

42. *Douglas Haig*, ed. Blake, 348–9.

43. Ferro, *Great War*, 226. M. Eksteins, *Rites of Spring: The Great War and the Birth of the Modern Age* (1989), 255–6. S. Hynes, *A War Imagined: The First World War and English Culture* (1990), 278–80.

44. Watt, *History*, 331. Stevenson, *First World War*, 310.

45. Eksteins, *Rites*, 256. Several authors of best-selling memoirs of the First World War, including Robert Graves and Charles Carrington (Charles Edmonds (pseud.), *A Subaltern's War*, Peter Davies, 1929), have explained that it took a decade or so before they could bear to relive their experiences in order to write about them.

46. Hynes, *War Imagined*, 119. This understandable rage on the part of combatants with false propaganda and the shortcomings of the home front is often misleadingly associated with 'anti-war' attitudes and the literature of 'disenchantment'. See Robert Graves, *Goodbye to All That* (Cassell, 1957) for a sensitive officer's revulsion at low morality and behaviour in England. Paul Fussell in *Wartime* (1989) rages against deception and betrayal of American troops by the deceit and propaganda on the home front in the Second World War.

47. R. Wohl, *The Generation of 1914* (1980), 219–22. For an eloquent statement of war's enduring fascination see Guy Chapman, *A Passionate Prodigality* (MacGibbon & Kee, 1965), 226.

48. Jonathan Marwil, *Frederic Manning: An Unfulfilled Life* (Durham, NC: Duke University Press, 1988), 179. In 1917 Manning wrote to William Rothenstein: 'I found that I felt most free in precisely those conditions when freedom seems to the normal mind least possible—an extraordinary feeling of self-reliance and self-assertion.'

49. Wohl, *Generation*, 223–5. Eksteins, *Rites*, 298. In the Prologue to the 2nd edn. of his war memoirs in 1957, Robert Graves made it clear that he had not attempted to write objectively. The book was hastily composed during a complicated domestic crisis when he desperately needed to make money. 'It was my bitter leave-taking of England where I had recently broken a good many conventions; quarrelled with, or been disowned by, most of my friends; been grilled by the police on suspicion of attempted murder; and ceased to care what anyone thought of me.' A few years later his proffered contribution to the regimental history of the Royal Welch Fusiliers was rejected on the grounds that there was too much fiction in it—see J. C. Dunn, *The War the Infantry Knew*, ed. Keith Simpson (Jane's, 1987), pp. xxx–xxxii.

50. Nicholas Hiley, 'Reality Reordered', *Times Literary Supplement* (16–22 Sept. 1988). B. Bond, *War and Society* (1984), 133–4. In 1920 Charles à Court Repington had provoked criticism by publishing a 2-vol. work entitled *The First World War*, since it was held to imply that there could be a Second. He could

equally well be criticized for suggesting that the conflict of 1914–18 had been the First World War.

51. See esp. *Paris or the Future of War* (Kegan Paul, 1925), *The Decisive Wars of History* (Bell, 1929), and *The Ghost of Napoleon* (New Haven: Yale Univ. Press, 1934). Also B. Bond, *Liddell Hart: A Study of his Military Thought* (1977), 42–53, and Christopher Bassford, *Clausewitz in English* (1994), 128–43.

52. Holsti, *Peace and War*, 228.

53. M. Howard, *War in European History* (1976), 118–19.

54. Notably J. S. Corum, *The Roots of Blitzkrieg: Hans von Seeckt and German Military Reform* (1992). See also L. M. Crowell, The Illusion of the Decisive Napoleonic Victory', *Defense Analysis*, 4/4 (Dec. 1988), 341–2.

55. Corum, *Roots*, 66, 201–2.

56. Howard, *Studies*, 108–9.

57. Bond, *Liddell Hart*, 54 ff. B. H. Reid, *J. F. C. Fuller: Military Thinker* (1987), 153, 159–60.

58. See e.g. I. F. Clarke, *Voices Prophesying War* (1970), 162 ff., and Barry D. Powers, *Strategy Without Slide-Rule: British Air Strategy, 1914–39* (Croom Helm, 1976).

59. Liddell Hart, *Paris*, 22, 43–7, 92. Bond, *Liddell Hart*, 38–42. I am grateful to Brian Holden Reid for the suggestion that Liddell Hart's ideas and language about strategic bombing in 1925 were greatly influenced by J. F. C. Fuller's *The Reformation of War* (1923).

60. B. Bond, *British Military Policy between the Two World Wars*, esp. chs. 7–9.

61. Wilson, *Myriad Faces*, 848–9.

62. Ibid. 853.

Chapter 7

1. P. M. H. Bell, *The Origins of the Second World War in Europe* (1986), 296–7.

2. John Lukacs, *The Last European War, September 1939–December 1941* (1977).

3. G. Blainey, *The Causes of War* (1973), 227. J. F. C. Fuller had previously stressed this point in *The Conduct of War, 1789–1961* (1972).

4. Michael Howard, *War and the Liberal Conscience* (1978), 110–14.

5. On cynicism about the American soldiers' lack of understanding of what the war was about see Paul Fussell, *Wartime* (1989). On Churchill's pursuit of victory for its own sake see Fuller, *Conduct of War*, 255–309.

6. Howard, *Liberal Conscience*, 114.

7. Bell, *Origins*, 164, 203, 298. Howard, *Liberal Conscience*, 101.

8. Bell, *Origins*, 198–9.

9. Ibid. 196–7. Bell provides an excellent summary on the differing views as to whether or not Hitler was planning a long and 'total' war.

10. See e.g. Bond, *Liddell Hart: A Study of his Military Thought* (1977) and John Mearsheimer, *Liddell Hart and the Weight of History* (Brassey's (UK) Ltd., 1988).

11. R. H. S. Stolfi, 'Equipment for Victory in France in 1940', *History* (Feb. 1970).

12. Williamson Murray, *The Change in the European Balance of Power, 1938–1939* (1984).

13. L. M. Crowell, 'The Illusion of the Decisive Napoleonic Victory', *Defense Analysis*, 4/4 (Dec. 1988), 343.

14. Telford Taylor, *The Breaking Wave: The German Defeat in the Summer of 1940* (Weidenfeld & Nicolson, 1967). John Lukacs, *The Duel: Hitler vs Churchill 10 May–31 July 1940* (1990).

15. See Paul Addison, 'Lloyd George and Compromise Peace in the Second World War', in A. J. P. Taylor (ed.), *Lloyd George: Twelve Essays* (Hamish Hamilton, 1971), 361–84.

16. Winston Churchill, *The Second World War* (Cassell, 1949), ii. 24.

17. Lukacs, *Duel*, 110. David Reynolds, 'Churchill in 1940: The Worst and Finest Hour', in Robert Blake and W. R. Louis (eds.), *Churchill* (OUP, 1993), 241–55.

18. Peter Calvocoressi, Guy Wint, and John Pritchard, *Total War: The Causes and Courses of the Second World War* (revised 2nd edn. 1989), 241.

19. Lukacs, *Last European War*, 187. For a map of 'Flights and Expulsions' see Calvocoressi *et al.*, *Total War*, 241.

20. Peter J. Liberman, 'Does Conquest Pay?', *Breakthroughs*, ii/2 (Cambridge, Mass.: MIT, Winter 1992/3), 6–11. This is an abstract of the author's Ph.D. thesis completed at MIT in Feb. 1992.

21. Calvocoressi *et al.*, *Total War*, 272–3.

22. Ibid. 275–8. Of the 5½ million prisoners captured by Germany on the Eastern Front about 3½ million were killed or died either in transit or in the camps.

23. Liberman, 'Conquest', 7.

24. Calvocoressi *et al.*, *Total War*, 282.

25. Liberman, 'Conquest', 11 n. 9.

26. Akira Iriye, *The Origins of the Second World War in Asia and the Pacific* (1989), 146–8.

27. Ibid. 150.

28. Ibid. 168–77. On the folly of the Japanese decision to attack Pearl Harbor see also J. F. C. Fuller, *The Second World War, 1939–1945* (Eyre & Spottiswoode, 1948), 127–34.

29. Martin Kitchen, *A World in Flames: A Short History of the Second World War in Europe and Asia, 1939–45* (1990), 151–3. See also Calvocoressi *et al.*, *Total War*, chs. 15–17.

30. Ibid. 997.

31. Kitchen, *World in Flames*, 152, 327–9.

32. Kitchen, *World in Flames*, 326–8. Liberman, 'Conquest', 11 n. 18.

33. Bell, *Origins*, 276–8.

34. P. M. H. Bell, *'A Certain Eventuality': Britain and the Fall of France* (Saxon House, 1974); Andrew Roberts, *'The Holy Fox': A Biography of Lord Halifax* (Weidenfeld & Nicolson, 1991).

35. Calvocoressi *et al.*, *Total War*, 422. See also Howard, *Liberal Conscience*, 109: 'It became accepted as much on the Left as on the Right that this was a war specifically against *Germany*, against a philosophy which appeared to be distinctively German and one which inspired a depressingly high proportion of the German people.' For the continuing controversy about Allied non-co-operation with the resisters see K. von Klemperer, *German Resistance against Hitler: The Search for Allies Abroad 1938–45* (OUP, 1992).

36. Hitler viewed the invasion of the Soviet Union more as an ideological crusade than as a problem of military strategy—hence in large part the operational muddle that occurred. See the discussion and reference in Bell, *Origins*, 294.

37. R. H. S. Stolfi, *Hitler's Panzers East: World War II Reinterpreted* (Alan Sutton: Far Thrupp, Stroud, 1992).

38. For a recent discussion of this issue see Nicholas Henderson, 'Hitler's Biggest Blunder', *History Today* (Apr. 1993), 35–43. Hitler was still confident about the outcome of the war so perhaps he was not being quixotic regarding Japan. He had a very poor opinion of Roosevelt and the USA.

39. Calvocoressi *et al.*, *Total War*, 422–3. John Erickson, *The Road to Berlin* (Boulder, Colo.: Westview Press, 1983), 162.

40. Kitchen, *World in Flames*, 243.

41. Sebastian Haffner, *The Meaning of Hitler* (Weidenfeld & Nicolson, 1979). B. Bond, *War and Society in Europe, 1870–1970* (1984), 184.

42. Quoted as the epigraph in Kitchen, *World in Flames*. Churchill's defiant rhetoric in 1940 ('We shall fight them on the beaches . . . we shall never surrender') may seem very similar to Hitler's, but Churchill was a patriot, a democrat, and a Parliamentarian. Had the worst nightmare come to pass in 1940 he would surely have surrendered power to someone who *was* prepared to negotiate— see the reference in n. 15 above.

43. For a robust British infantryman's view of the Japanese as enemies, all the more impressive because it challenges more recent interpretations of what the war in the East was really like, see George MacDonald Fraser, *Quartered Safe Out Here: A Recollection of the War in Burma* (Harvill, 1992).

44. Kitchen, *World in Flames*, 329. Calvocoressi *et al.*, *Total War*, 1163–4.

45. Kitchen, *World in Flames*, 330.

46. Michael Balfour, 'The Origin of the Formula "Unconditional Surrender" in World War II', *Armed Forces and Society*, 5/2 (Feb. 1979), 281–301, and the same author's *Propaganda in War, 1939–1945* (1979), 314–20, 365–7.

Balfour's seemingly incontrovertible arguments have been overlooked by critics who apparently want to believe that 'unconditional surrender' was a wicked policy which stiffened German resistance and unnecessarily prolonged the war. See e.g. Richard Hobbs, *The Myth of Victory: What is Victory in War?* (Boulder, Colo.: Westview Press, 1979), esp. part 2, pp. 135–264.

47. Quoted in Balfour, *Propaganda*, 367.
48. Bond, *War and Society*, 196.
49. Calvocoressi *et al.*, *Total War*, 422–5. For a robust treatment of this controversy see G. L. Weinberg, *A World at Arms* (1994), 438–9, 482–3, 798.
50. Quoted by Gordon A. Craig in 'Good Germans', *New York Review of Books* (17 Dec. 1992). See also Balfour, *Propaganda*, 318.
51. Ibid. 320.
52. See e.g. Michael Howard, *The Mediterranean Strategy in the Second World War* (Weidenfeld & Nicolson, 1968) or, for a more critical view T. Ben-Moshe, *Churchill: Strategy and History* (Hemel Hempstead: Harvester Wheatsheaf, 1992).
53. Bond, *War and Society*, 191–3. R. J. Overy, *The Air War, 1939–1945* (1980). Ian Kershaw, *The 'Hitler Myth': Image and Reality in the Third Reich* (1989).
54. Overy, *Air War*, 121–3. Kershaw, *Hitler Myth*, 200–7. See also Weinberg, *World at Arms*, 579–81.
55. Overy, *Air War*, 121–3.
56. Ibid. 123–5.
57. Charles Webster and Noble Frankland, *The Strategic Air Offensive against Germany* (HMSO, 1961), iii. 112, and see p. 108 for previous air attacks on Dresden.
58. Sir Arthur Harris, *Bomber Offensive* (Collins, 1947), 261.
59. Webster and Frankland, *Strategic Offensive*, 52–5. See also F. H. Hinsley, *British Intelligence in the Second World War*, iii/2 (HMSO, 1988), 608–13, 624. The V2 rocket attacks on London ceased abruptly on 27 Mar. 1945.
60. Webster and Frankland, *Strategic Offensive*, 117.
61. For one bomber crew member who later became preoccupied with the Dresden raid because the city was known to be packed with refugees, see Miles Tripp, *The Eighth Passenger* (Macmillan Paperback, 1985), 87–9, 171–2, 177.
62. Ian Buruma, 'The Devils of Hiroshima', *New York Review of Books* (25 Oct. 1990). Buruma notes the enigmatic inscription on the Hiroshima Cenotaph by one Professor Saika, 'Please rest in peace, for the error will not be repeated'. But *whose* error is unclear. The Japanese have been educated to regard the bomb as a blinding bolt from a sunny sky delivered by a demonized America. At last, however, under a new emperor, in 1993, 'Japan Prepares to Repent its War Crimes', report in *Daily Telegraph* (13 Aug. 1993).
63. See the excellent accounts in Calvocoressi *et al.*, *Total War*, 1163 ff., and

Lawrence Freedman, 'The Strategy of Hiroshma', *Journal of Strategic Studies*, 1/1 (May 1978), 76–97.

64. Kitchen, *World in Flames*, 330–3. For a recent view that American policy-makers and service leaders did not anticipate a sudden end of the war, even after the use of atomic bombs, see Jacob Vander Meulen, 'Planning for V-J Day by the U.S. Army Air Forces and the Atomic Bomb Controversy', *Journal of Strategic Studies*, 16/2 (June 1993), 227–39.

65. Kitchen, *World in Flames*, 334–5. For the deep cultural roots of the Japanese preference for suicide rather than surrender see Ivan Morris, *The Nobility of Failure: Tragic Heroes in the History of Japan* (Secker & Warburg, 1975).

66. W. Churchill, *The Second World War* (Cassell, 1951), iv. 553.

67. Laurens van der Post, *The Night of the New Moon* (Hogarth Press, 1970). See also Leonard Cheshire, *The Light of Many Suns: The Meaning of the Bomb* (Methuen, 1985), 62, 65. Aidan MacCarthy, *A Doctor's War* (Robson Paperback, 1985), 124 ff. See also the robust argument in defence of the dropping of the bomb in George MacDonald Fraser, *Quartered Safe*, 216–21.

68. Calvocoressi *et al.*, *Total War*, 1188. On the continuing problems of the Japanese in coming to terms with their role in the war see Ian Buruma, *Wages of Guilt: Memories of the War in Germany and Japan* (Cape, 1994).

69. F. H. Hinsley, 'British Intelligence in the Second World War', in C. Andrew and J. Noakes (eds.), *Intelligence and International Relations 1900–1945* (Exeter: Univ. of Exeter, 1987), 217–18.

70. Liddell Hart's views during the Second World War were idiosyncratic, to put it mildly, but he was quick to see the problems which would result from destroying Germany too completely. See Bond, *Liddell Hart*, ch. 5.

71. Liddell Hart, *Why don't we Learn from History?* (Allen & Unwin, 1972). In his postumously published *History of the Second World War* (Cassell, 1970), 713, Liddell Hart concluded that ' "the unnecessary war" was unnecessarily prolonged, and millions more lives needlessly sacrificed, while the ultimate peace merely produced a fresh menace and the looming fear of another war'. For extremely positive judgements on the significance of victory in the Second World War see Kitchen, *World in Flames*, 344; J. M. Roberts, *A General History of Europe 1880–1945* (1976), 540; and Calvocoressi *et al.*, *Total War*, 589–92.

72. A. J. P. Taylor, *The Second World War: An Illustrated History* (Hamish Hamilton, 1975), 234.

73. Roberts, *Europe*, 540.

74. Kitchen, *World in Flames*, 344. Calvocoressi *et al.*, *Total War*, 592. Weinberg, *World at Arms*, 898–9.

Chapter 8

1. J. Keegan, *The Face of Battle* (1976), 336.
2. C. Cook and J. Stevenson, *The Longman Handbook of World History since 1914* (1991 paperback edn.), 280–305.
3. A. Buchan, *War in Modern Society* (1966), 127–9.
4. Ibid. 40.
5. M. Van Creveld, *The Transformation of War* (1991), 154. The English edn., incorporating a few amendments, is titled *On Future War* (1991).
6. This does not contradict the point made in the previous chapter that territorial conquest in war can be made to yield economic benefits to a ruthless occupier.
7. See e.g. M. Handel (ed.), 'Clausewitz and Modern Strategy', *Journal of Strategic Studies*, 9 (June and Sept. 1986). John Keegan, *A History of Warfare* (1993).
8. M. Howard, *The Causes of Wars*, 140–1. Clausewitz, *On War*, ed. M. Howard and P. Paret (1976), 605–10, 'War is an Instrument of Policy'.
9. Howard, *Causes*, 147, and *Clausewitz* (1983), 71.
10. Ibid. 51, 69–73.
11. Howard, *Causes*, 136–7, 141.
12. Colin S. Gray, *War, Peace and Victory: Strategy and Statecraft for the Next Century* (New York: Simon & Schuster, Touchstone paperback, 1991), 260. Howard defends Brodie's view that there could be no winner in a nuclear war against Gray's criticisms in *Causes*, 133–50.
13. Van Creveld, *Transformation*, 36, 58–62, 192.
14. Ibid. 194. M. Howard, 'Famous Last Screams', *London Review of Books* (5 Dec. 1991).
15. Van Creveld, *Transformation*, 176–8.
16. Ibid. 124, 142–9, 221–7.
17. Ibid. 211–12.
18. Howard, *Causes*, 89–90. John Shy and Thomas W. Collier, 'Revolutionary War', in P. Paret (ed.), *Makers of Modern Strategy* (1986), 815–62. See esp. 822, 860. This point is also made by John Mueller in *Retreat from Doomsday* (1989), 176.
19. Clausewitz, *On War*, 370.
20. Buchan, *Modern Society*, 80–3, 92–3.
21. M. Carver, *War since 1945* (1980), 165–6.
22. M. Hastings, *The Korean War* (Pan paperback, 1988), 422–3.
23. Ibid. 414, 419. In an interview with Trevor Macdonald for ITV on 7 Aug. 1993 General Colin Powell, Chairman of the US Joint Chiefs of Staff repeatedly stressed that if the USA carried out air strikes in Bosnia they would be 'decisive'. It is difficult to grasp what this word could mean in the circumstances.

24. Carver, *War*, 203–33.

25. Chaim Herzog, *The Arab–Israeli Wars: War and Peace in the Middle East* (Arms & Armour Press, 1982). Carver, *War*, 234–72.

26. Herzog, *Arab–Israeli Wars*, 145–91. J. Keegan, 'The Six Day Miracle', *Daily Telegraph* (6 June 1992).

27. A. Hertzberg, 'Israel: The Tragedy of Victory', *New York Review of Books* (28 May 1987).

28. Shy and Collier, 'Revolutionary War', 855 ff.

29. Gray, *War, Peace, Victory*, 115–16. Hastings, *Korean War*, 420. John Guilmartin, 'Vietnam lessons in Bosnia', letter in the *Daily Telegraph* (13 Aug. 1993). In *Trapped by Success: The Eisenhower Administration and Vietnam, 1953–61* (Columbia UP, 1991) David L. Anderson shows that the USA did not carefully consider the prospects of success before intervening. I am indebted to Dr Brian Holden Reid for this reference.

30. Cook and Stevenson, *Longman Handbook*, 115–17.

31. Guenter Lewy, *America in Vietnam* (1978), 32–41, 424. M. Van Creveld, *Command in War* (1985), 232–60. Mueller, *Retreat*, 177–80.

32. Robert Elegant, 'Vietnam: How to Lose a War', *Encounter* (Aug. 1981), 73–90. The author was a foreign correspondent in Asia, 1951–76, and spent extensive periods in Vietnam covering the war for *Newsweek* and other US news services. In the years concerned he received three Overseas Press Club Awards. See also Lewy, *America*, 307–42.

33. Hastings, *Korean War*, 413.

34. Jonathan Mirsky, 'Reconsidering Vietnam', *New York Review of Books* (10 Oct. 1991). After 19 years the USA's trade embargo on Vietnam was lifted on 4 Feb. 1994.

35. Julian Pettifer, 'Frontline' television Channel 4 (27 June 1993) and review in the *Daily Telegraph* (28 June 1993). See also Hastings, *Korean War*, 426. As soon after the war as 1978 Lewy (*America*, 435) wrote 'Shrill rhetoric created a world of unreality in which the North Vietnamese Communists were the defenders of national self-determination, while U.S. actions designed to prevent the forceful takeover of South Vietnam stood branded as imperialism and aggression'.

36. E. Luttwak reviewing Admiral Sandy Woodward's memoirs, *One Hundred Days*, and reported in the *Sunday Telegraph* on 3 May 1992 under the heading 'How we Lost the Falklands War'.

37. M. E. Yapp, *The Near East since the First World War* (Longman Paperback, 1991), 427–32. E. Goldstein, *Wars and Peace Treaties 1816–1991* (1992), 133–5.

38. L. Freedman and E. Karsh, *The Gulf Conflict 1990–1991: Diplomacy and War in the New World Order* (1993), 403–8, and review 'What New World Order?'

by Patrick Glynn, *Times Literary Supplement* (30 Apr. 1993). 'Iraq Won the War, Says Saddam', *Daily Telegraph* (30 July 1991).

39. Freedman and Karsh, *Gulf Conflict*, 412–18.

40. Mueller, *Retreat*, 217–19, 236–9, 253–65, concludes that inter-state wars within the developed world are obsolescent and irrational but concedes in theory that if they should occur they might be kept under control and limited. Writing in the late 1980s and concentrating almost exclusively on 'the first world', he none the less expressed optimism about the likely decline of warfare in 'the third world'. As a long-term prophecy this may still prove accurate but at present (early 1994) the prospects remain grim.

41. Freedman and Karsh, *Gulf Conflict*, 44. On Mueller's optimistic forecast see also Carl Kaysen, 'Is War Obsolete? A Review Essay', *International Security*, 14/4 (Spring 1990), 42–64.

Conclusion

1. Justin Wintle (ed.), *The Dictionary of War Quotations* (New York: The Free Press, 1989), 91.

2. Edward N. Luttwak, *On the Meaning of Victory* (1986), 289–93.

3. I was born in 1936. I *believe* my earliest memory of a public event was the successful action of the cruisers *Exeter*, *Ajax*, and *Achilles* against the pocket-battleship *Graf Spee* on 13 Dec. 1939. I certainly remember the sight of London burning from the Chiltern Hills, nearly 40 miles away.

4. Paul W. Schroeder, *The Transformation of European Politics, 1763–1848* (1994), 390–5.

SELECT BIBLIOGRAPHY

Publication details of books cited only once are given in the references. London is the place of publication unless otherwise stated.

Balfour, Michael, *Propaganda in War, 1939–1945* (Routledge & Kegan Paul, 1979).

Barclay, C. N. *Armistice, 1918* (J. M. Dent, Aldine Paperback, 1969).

Bassford, Christopher, *Clausewitz in English: The Reception of Clausewitz in Britain and America, 1815–1945* (New York: OUP, 1994).

Becker, Jean-Jacques, *The Great War and the French People* (Leamington Spa: Berg, 1985).

Bell, P. M. H., *The Origins of the Second World War in Europe* (Longman, 1986).

Berghahn, Volker R. and Kitchen, Martin (eds.), *Germany in the Age of Total War* (Croom Helm, 1981).

Bernhardi, Friedrich von, *Germany and the Next War* (Arnold, 1912).

Bessel, Richard, *Germany after the First World War* (Oxford: Clarendon Press, 1993).

Blainey, Geoffrey, *The Causes of War* (Macmillan, 1973).

Bloch, Ivan S., *Is War Impossible?* (Gregg Revivals, 1991).

Bond, Brian, *Liddell Hart: A Study of his Military Thought* (Cassell, 1977, and Gregg Revivals, 1991).

—— *War and Society in Europe, 1870–1970* (Collins, Fontana Paperback, 1984).

Buchan, Alastair, *War in Modern Society* (C. A. Watts, 1966).

Bucholz, Arden, *Moltke, Schlieffen and Prussian War Planning* (Leamington Spa: Berg, 1991).

Calvocoressi, Peter, Wint, Guy, and Pritchard, John, *Total War: The Causes and Courses of the Second World War* (revised 2nd edn. Viking, 1989).

Carr, William, *The Origins of the Wars of German Unification* (Longman, 1991).

Carver, Michael, *War since 1945* (Weidenfeld & Nicolson, 1980).

Clarke, I. F., *Voices Prophesying War, 1763–1984* (Panther Paperback, 1970).

Clausewitz, Carl von, *On War*, ed. Michael Howard and Peter Paret (Princeton: Princeton UP, 1976).

Colin, Jean, *The Transformations of War* (Hugh Rees, 1912).

Corum, James S., *The Roots of Blitzkrieg: Hans von Seeckt and German Military Reform* (Lawrence, Kan.: UP of Kansas, 1992).

Craig, Gordon A., *The Politics of the Prussian Army, 1640–1945* (New York: OUP, Galaxy Paperback, 1964).

Craig, Gordon A., *The Battle of Königgrätz* (Weidenfeld & Nicolson, 1965).

—— *Germany: 1866–1945* (Oxford: Clarendon Press, 1978).

Duffy, Christopher, *The Army of Frederick the Great* (Newton Abbot: David & Charles, 1974).

—— *Frederick the Great: A Military Life* (Routledge & Kegan Paul, 1985).

Eksteins, Modris, *Rites of Spring: The Great War and the Birth of the Modern Age* (Bantam Press, 1989).

Evans, R. J. W., and Strandmann, H. P. von (eds.), *The Coming of the First World War* (Oxford: Clarendon Press, 1988).

Ferro, Marc, *The Great War, 1914–18* (Routledge & Kegan Paul, 1973).

Freedman, Lawrence, and Karsh, Efraim, *The Gulf Conflict 1990–1: Diplomacy and War in the New World Order* (Faber, 1993).

Fuller, J. F. C., *The Conduct of War, 1789–1961* (Eyre, Methuen, University Paperback, 1972).

Fussell, Paul, *Wartime: Understanding and Behavior in the Second World War* (New York: OUP, 1989).

Gallie, W. B., *Philosophers of Peace and War* (Cambridge: CUP, 1978).

Gat, Azar, *The Origins of Military Thought from the Enlightenment to Clausewitz* (Oxford: Clarendon Press, 1989).

—— *The Development of Military Thought: The Nineteenth Century* (Oxford: Clarendon Press, 1992).

Goldstein, Erik, *Wars and Peace Treaties 1816–1991* (Routledge, 1992).

Haig, Douglas, *The Private Papers of Douglas Haig, 1914–1919*, ed. Robert Blake (Eyre & Spottiswoode, 1952).

Hastings, Max, *The Korean War* (Pan Paperback, 1988).

Herzog, Chaim, *The Arab-Israeli Wars: War and Peace in the Middle East* (Arms & Armour Press, 1982).

Hinsley, F. H., *Power and the Pursuit of Peace* (Cambridge: CUP, 1963).

Holsti, Kalevi J., *Peace and War: Armed Conflicts and International Order, 1648–1989* (Cambridge: CUP, 1991).

Howard, Michael, *The Franco-Prussian War* (Collins, Fontana, 1967).

—— *Studies in War and Peace* (Temple Smith, 1970).

—— *War in European History* (Oxford: OUP, 1976).

—— *War and the Liberal Conscience* (Temple Smith, 1978).

—— *Clausewitz* (Oxford: OUP, 1983).

—— *The Causes of Wars* (Temple Smith, 1983).

Hynes, Samuel, *A War Imagined: The First World War and English Culture* (The Bodley Head, 1990).

Iriye, Akira, *The Origins of the Second World War in Asia and the Pacific* (Longman, 1989).

Joll, James, *Europe since 1870* (Penguin Books, 1976).

Jomini, Antoine-Henri de, *The Art of War* (Greenhill Books, 1992).

Kaiser, David, *Politics and War: European Conflict from Philip II to Hitler* (I. B. Tauris, 1990).

Keegan, John, *The Face of Battle* (Cape, 1976).

—— *A History of Warfare* (Hutchinson, 1993).

Kennedy, Paul (ed.), *Grand Strategies in War and Peace* (New Haven: Yale UP, 1991).

Kershaw, Ian, *The 'Hitler Myth': Image and Reality in the Third Reich* (OUP, paperback edn., 1989).

Kitchen, Martin, *The Silent Dictatorship: The Politics of the German High Command under Hindenburg and Ludendorff* (Croom Helm, 1976).

—— *A World in Flames: A Short History of the Second World War in Europe and Asia, 1939–45* (Longman, 1990).

Lewy, Guenter, *America in Vietnam* (New York: OUP, 1978).

Liddell Hart, Basil H., *The Ghost of Napoleon* (New Haven: Yale UP, 1934).

Lukacs, John, *The Last European War, September 1939–December 1941* (Routledge & Kegan Paul, 1977).

—— *The Duel: Hitler vs Churchill 10 May–31 July 1940* (The Bodley Head, 1990).

Luttwak, Edward N., *On the Meaning of Victory* (New York: Simon & Schuster, 1986).

Luvaas, Jay, *The Military Legacy of the Civil War: The European Inheritance* (Chicago: Chicago UP, 1959).

—— (ed.), *Frederick the Great on the Art of War* (New York: The Free Press, 1966).

Mueller, John, *Retreat from Doomsday: The Obsolescence of Major War* (New York: Basic Books, 1989).

Murray, Williamson, *The Change in the European Balance of Power, 1938–1939: The Path to Ruin* (Princeton: Princeton UP, 1984).

Overy, R. J., *The Air War, 1939–1945* (Europa Publications, 1980).

Paret, Peter, *Clausewitz and the State* (Princeton: Princeton UP, 1985).

—— (ed.), *Makers of Modern Strategy from Machiavelli to the Nuclear Age* (Princeton: Princeton UP, 1986).

Porch, Douglas, *The March to the Marne: The French Army 1871–1914* (Cambridge: CUP, 1981).

Prior, Robin, and Wilson, Trevor, *Command on the Western Front: The Military Career of Sir Henry Rawlinson 1914–1918* (Oxford: Blackwell, 1992).

Reid, Brian Holden, *J. F. C. Fuller: Military Thinker* (Macmillan, 1987).

Ritter, Gerhard, *The Schlieffen Plan: Critique of a Myth* (Wolff, 1958).

—— *Frederick the Great* (Eyre & Spottiswoode, 1968).

—— *The Sword and the Scepter: The Problem of Militarism in Germany*, 4 vols. (Coral Gables, Fl.: University of Miami Press, 1969–73).

Roberts, J. M., *A General History of Europe 1880–1945* (Longman, 1976).

Rothenberg, Gunther E., *The Art of Warfare in the Age of Napoleon* (Batsford, 1977).

Schroeder, Paul W., *The Transformation of European Politics, 1763–1848* (Oxford: Clarendon Press, 1994).

Sharp, Alan, *The Versailles Settlement: Peacemaking in Paris, 1919* (Macmillan, 1991).

Stevenson, David, *The First World War and International Politics* (Oxford: OUP, 1988).

Strachan, Hew, *European Armies and the Conduct of War* (Allen & Unwin, 1983).

Taylor, A. J. P., *The Struggle for Mastery in Europe, 1848–1918* (Oxford: Clarendon Press, 1969).

Travers, Tim, *The Killing Ground: The British Army, the Western Front and the Emergence of Modern Warfare 1900–1918* (Allen & Unwin, 1987).

—— *How the War was Won: Command and Technology in the British Army on the Western Front, 1917–1918* (Routledge, 1992).

Van Creveld, Martin, *Supplying War: Logistics from Wallenstein to Patton* (Cambridge: Cambridge UP, 1978).

—— *Command in War* (Cambridge, Mass.: Harvard UP, 1985).

—— *The Transformation of War* (New York: Free Press, 1991; UK edn., *On Future War*, Brassey's, 1991).

Von der Goltz, Colmar, *The Nation in Arms* (Hugh Rees, 1903).

Watt, Donald Cameron, *A History of the World in the 20th Century*, i. *1899–1918* (Pan Books, 1970).

—— *Too Serious a Business: European Armed Forces and the Approach to the Second World War* (Temple Smith, 1975).

Webster, Charles, and Frankland, Noble, *The Strategic Air Offensive against Germany*, iii (HMSO, 1961).

Weigley, Russell F., *The Age of Battles: The Quest for Decisive Warfare from Breitenfeld to Waterloo* (Bloomington, Ind.: Indiana UP, 1991).

Weinberg, Gerhard L., *A World at Arms. A Global History of World War II* (Cambridge: Cambridge UP, 1994).

Weintraub, Stanley, *A Stillness Heard around the World: The End of the Great War, November 1918* (Oxford: OUP, 1985).

Wilkinson, Spenser, *The French Army before Napoleon* (Gregg Revivals, 1991).

—— *The Rise of General Bonaparte* (Gregg Revivals, 1991).

Wilson, Henry, *Field Marshal Sir Henry Wilson: His Life and Diaries*, ed. Charles E. Callwell, 2 vols. (Cassell, 1927).

Wilson, Trevor, *The Myriad Faces of War: Britain and the Great War, 1914–1918* (Oxford: Blackwell, Polity Press, 1986).

Wohl, Robert, *The Generation of 1914* (Weidenfeld & Nicolson, 1980).

INDEX